NISTIR 7955

I0488466

Dietary Supplement Laboratory Quality Assurance Program: Exercise I Final Report

Melissa M. Phillips
Catherine A. Rimmer
Laura J. Wood

Karen E. Murphy
Thomas W. Vetter
Chemical Sciences Division
Material Measurement Laboratory

August 2013

U.S. Department of Commerce
Penny Pritzker, Secretary

National Institute of Standards and Technology
Patrick D. Gallagher, Under Secretary of Commerce for Standards and Technology and Director

TABLE OF CONTENTS

CATECHINS IN GREEN TEA

ABSTRACT

The NIST Dietary Supplement Laboratory Quality Assurance Program (DSQAP) was established in collaboration with the National Institutes of Health (NIH) Office of Dietary Supplements (ODS) in 2007 to enable members of the dietary supplements community to improve the accuracy of measurements for demonstration of compliance with various regulations. Exercise I of this program offered the opportunity for laboratories to assess their in-house measurements of nutritional elements (Cr, Mo, and Se), contaminants (Cd), water-soluble vitamins (pantothenic acid), fat-soluble vitamins (retinol), and catechins in foods and/or botanical dietary supplement ingredients and finished products.

INTRODUCTION

The dietary supplement industry in the US is booming, with two-thirds of adults considering themselves to be supplement users.[1] Consumption of dietary supplements, which includes vitamin and mineral supplements, represents an annual US expenditure of more than $25 billion. These figures represent an increasing American trend, and as a result, it is critically important that both the quality and safety of these products are verified and maintained.

The Dietary Supplement Health and Education Act of 1994 (DSHEA) amended the Food, Drug and Cosmetic Act to create the regulatory category called dietary supplements. The DSHEA also gave the FDA authority to write current Good Manufacturing Practices (cGMPs) that require manufacturers to evaluate the identity, purity, and composition of their ingredients and finished products. To enable members of the dietary supplements community to improve the accuracy of the measurements required for compliance with these and other regulations, NIST established the Dietary Supplement Laboratory Quality Assurance Program (DSQAP) in collaboration with the NIH ODS in 2007.

The program offers the opportunity for laboratories to assess their in-house measurements of active or marker compounds, nutritional elements, contaminants (toxic elements, pesticides, mycotoxins), and fat- and water-soluble vitamins in foods as well as in botanical dietary supplement ingredients and finished products. Reports and certificates of participation are provided and can be used to demonstrate compliance with the cGMPs. In addition, NIST and the DSQAP assist the ODS Analytical Methods and Reference Materials program (AMRM) at the NIH in supporting the development and dissemination of analytical tools and reference materials. In the future, results from DSQAP exercises could be used by ODS to identify problematic matrices and analytes for which an Official Method of Analysis would benefit the dietary supplement community.

NIST has experience in the area of quality assurance programs, but the DSQAP takes a unique approach. In other NIST quality assurance programs, a set of analytes is measured repeatedly over time in the same or similar matrices to demonstrate laboratory performance. In contrast, the wide range of matrices and analytes under the "dietary supplement" umbrella means that not

---

[1] Walsh, T. (2012) *Supplement Usage, Consumer Confidence Remain Steady According to New Annual Survey from CRN.* Council for Responsible Nutrition, Washington, DC.

every laboratory is interested in every sample or analyte. The constantly changing dietary supplement market, and the enormous diversity of finished products, makes repeated determination of a few target compounds in a single matrix of little use to participants. Instead, participating laboratories are interested in testing in-house methods on a wide variety of challenging, real-world matrices to demonstrate that their performance is comparable to that of the community and that their methods provide accurate results. In an area where there are few standard methods, the DSQAP offers a unique tool for assessment of the quality of measurements, provides feedback about performance, and can assist participants in improving laboratory operations.

This report summarizes the results from the ninth exercise of the DSQAP, Exercise I. Eighty-five laboratories responded to the call for participants distributed in October 2012. Samples were shipped to participants in December 2012, and results were returned to NIST by March 2013. This report contains the final data and information to be disseminated to the participants in July 2013.

OVERVIEW OF DATA TREATMENT AND REPRESENTATION

Statistics

The individual data table and graphs contain information about the performance of each laboratory relative to that of the other participants in this study and relative to a target around the expected result (if available). The consensus mean and standard deviation are calculated according to the robust algorithm outlined in ISO 13528:2005(E), Annex C.[2] The algorithm is summarized here in simplified form.

Initial values of the consensus mean, x^*, and consensus standard deviation, s^*, are estimated as

$$x^* = \text{median of } x_i \qquad (i = 1, 2, \ldots, n)$$
$$s^* = 1.483 \times \text{median of } |x_i - x^*| \qquad (i = 1, 2, \ldots, n).$$

These initial values for x^* and s^* are updated by first calculating the expanded standard deviation, δ, as

$$\delta = 1.5 \times s^*.$$

Then each x_i is compared to the expanded range and adjusted to x_i^* as described below to reduce the effect of outliers.

$$\text{If } x_i < x^* - \delta, \text{ then } x_i^* = x^* - \delta.$$
$$\text{If } x_i > x^* + \delta, \text{ then } x_i^* = x^* + \delta.$$
$$\text{Otherwise, } x_i^* = x_i.$$

New values of x^*, s^*, and δ are calculated iteratively until the process converges. Convergence is taken as no change from one iteration to the next in the third significant figure of s^* and in the equivalent digit in x^*:

$$x^* = \frac{\sum_{i=1}^{n} x_i^*}{n}$$

$$s^* = 1.134 \times \sqrt{\frac{\sum_{i=1}^{n}(x_i^* - x^*)}{n-1}}.$$

Individual Data Table

The data in this table is individualized to each participating laboratory and is provided to allow participants to directly compare their data to the summary statistics (consensus or community data as well as NIST certified, reference, or estimated values). The upper left of the data table includes the randomized laboratory code. Tables included in this report are generated using NIST data to protect the identity and performance of participants.

Section 1 of the data table contains the laboratory results as reported, including the mean and standard deviation when multiple values were reported. A blank indicates that NIST does not have data on file for that laboratory for a particular analyte or matrix. An empty box for standard deviation indicates that only a single value was reported and therefore that value was not included in the calculation of the consensus data.[2]

Also in Section 1 are two Z-scores. The first Z-score, Z_{comm}, is calculated with respect to the community consensus value, using x^* and s^*:

$$Z_{comm} = \frac{x_i - x^*}{s^*}.$$

The second Z-score, Z_{NIST}, is calculated with respect to the target value (NIST certified, reference, or estimated value), using x_{NIST} and U_{95} (the expanded uncertainty) or s_{NIST}, the standard deviation of NIST measurements:

$$Z_{NIST} = \frac{x_i - x_{NIST}}{U_{95}}$$

or

$$Z_{NIST} = \frac{x_i - x_{NIST}}{s_{NIST}}.$$

The significance of the Z-score is as follows:
- $|Z| < 2$ indicates that the laboratory result is considered to be within the community consensus range (for Z_{comm}) or NIST target range (for Z_{NIST}).
- $2 < |Z| < 3$ indicates that the laboratory result is considered to be marginally different from the community consensus value (for Z_{comm}) or NIST target value (for Z_{NIST}).
- $|Z| > 3$ indicates that the laboratory result is considered to be significantly different from the community consensus value (for Z_{comm}) or NIST target value (for Z_{NIST}).

---

[2] ISO 13528:2005(E), *Statistical methods for use in proficiency testing by interlaboratory comparisons*, pp 14-15.

Section 2 of the data table contains the community results, including the number of laboratories reporting more than a single value for a given analyte[1], the mean value determined for each analyte, and a robust estimate of the standard deviation of the reported values.[3] Consensus means and standard deviations are calculated using the laboratory means; if a laboratory reported a single value, the reported value is not included.[1] Additional information on calculation of the consensus mean and standard deviation can be found in the previous section.

Section 3 of the data table contains the target values for each analyte. When possible, the target value is a certified or reference value determined at NIST. Certified values and the associated expanded uncertainty (U_{95}) have been determined with two independent analytical methods at NIST, by collaborating laboratories, or in some combination. Reference values are assigned using NIST values obtained from the average and standard deviation of measurements made using a single analytical method. For both certified and reference values, at least six samples have been tested and duplicate preparations from the sample package have been included, allowing the uncertainty to encompass variability due to inhomogeneity within and between packages. For samples in which a NIST certified or reference value is not available, the analytes are measured at NIST using an appropriate method. The NIST-assessed value represents the mean of at least three replicates. For materials acquired from another proficiency testing program, the consensus value and uncertainty from the completed round is used as the target range.

Summary Data Table
This data table includes a summary of all reported data for a particular analyte in a particular study. Participants can compare the raw data for a single laboratory to data reported by the other participating laboratories or to the consensus data. A blank indicates that the laboratory signed up and received samples for that particular analyte and matrix, but NIST does not have data on file for that laboratory.

Graphs
Data Summary View (Method Comparison Data Summary View)
In this view, individual laboratory data are plotted with the individual laboratory standard deviation (error bars). Data points that are unfilled represent laboratories that only reported a single value for that analyte and therefore were not included in the consensus mean. The black solid line represents the consensus mean, and the black dotted lines represent the consensus variability calculated as one standard deviation about the consensus mean. The gray shaded region represents the target zone for "acceptable" performance, which encompasses the NIST certified, reference, or estimated value bounded by twice its uncertainty (U_{95}) or standard deviation. For the purpose of the DSQAP, a target range spanning twice the uncertainty in the NIST value is selected because participants are only asked to make a limited number of observations. The size of the y-axis on the data summary view graph represents the consensus mean bounded by 2δ. In this view, the relative locations of individual laboratory data and consensus zones with respect to the target zone can be compared easily. In most cases, the target zone and the consensus zone overlap, which is the expected result. One program goal is to reduce the size of the consensus zone and center the consensus zone about the target value.

---

[3] ISO 13528:2005(E), *Statistical methods for use in proficiency testing by interlaboratory comparisons*, Annex C.

Analysis of an appropriate reference material as part of a quality control scheme can help to identify sources of bias for laboratories reporting results that are significantly different from the target zone. In the case in which a method comparison is relevant, different colored data points may be used to indicate laboratories that used a specific approach to sample preparation, analytical method, or quantitation.

Sample/Control Comparison View (Sample/Sample Comparison View)
In this view, the individual laboratory results for a control (NIST SRM with a certified value) are compared to the results for an unknown (another NIST SRM with a more challenging matrix, a commercial sample, etc.). The error bars represent the individual laboratory standard deviation. The solid red box represents the target zone for the control (x-axis) and unknown sample (y-axis). The dotted blue box represents the consensus zone for the control (x-axis) and the unknown sample (y-axis). This view emphasizes trends in the data that may indicate potential calibration issues or method biases. One program goal is to identify such calibration or method biases and assist participants in improving analytical measurement capabilities. In some cases, when two equally challenging materials are provided, the same view (sample/sample comparison) can be helpful in identifying commonalities or differences in the analysis of the two materials.

TRACE NUTRITIONAL ELEMENTS IN FOODS AND SUPPLEMENTS

Study Overview
In this study, participants were provided with one NIST SRM, SRM 3280 Multivitamin/Multielement Tablets, and a powdered whole egg material. Participants were asked to use in-house analytical methods to determine the mass fractions of three nutritional elements (chromium, molybdenum, and selenium) in each of the matrices and report values on an as-received basis.

Sample Information
Multivitamin/multielement tablets. Participants were provided with one packet containing 15 multivitamin/multielement tablets. The material was produced by blending a vitamin and mineral pre-mix with a direct-compression tablet formulation. Intact tablets were heat-sealed inside 0.1 mm (4 mil) polyethylene bags, which were then sealed inside Mylar bags. Before use, participants were instructed to grind all tablets together, mix the resulting powder thoroughly, and use a sample size of at least 0.5 g. Participants were asked to store the material at controlled room temperature, 10 °C to 30 °C, prepare three samples, and report three values from the resulting ground material. Approximate analyte levels were not provided to participants prior to the study. NIST certified values in SRM 3280 were determined using inductively coupled plasma mass spectrometry (ICP-MS), inductively coupled plasma optical emission spectrometry (ICP-OES), instrumental neutron activation analysis (INAA), and X-ray florescence spectroscopy (XRF). The certified values and uncertainties for Cr, Mo, and Se in SRM 3280 are outlined in the table below, both on a dry-mass basis and an as-received basis following adjustment for the moisture content of the material (1.37 %).

Analyte	Certified Mass Fraction (mg/kg) (dry-mass basis)			Adjusted Mass Fraction (µg/g) (as-received basis)		
Cr	93.7	±	2.7	92.4	±	2.7
Mo	70.7	±	4.5	69.7	±	4.4
Se	17.42	±	0.45	17.2	±	0.4

Whole egg powder. Participants were provided with one packet containing approximately 10 g of commercially available whole egg powder. The whole egg powder is a free-flowing, fine powder prepared from USDA-inspected eggs. The powder was heat-sealed inside nitrogen-flushed 0.1 mm (4 mil) polyethylene bags, which were then sealed inside aluminized plastic bags. Before use, participants were instructed to thoroughly mix the contents of the packet and use a sample size of at least 0.5 g. Participants were asked to store the material at controlled room temperature, 10 °C to 30 °C and report three values from the single packet provided. Approximate analyte levels were not provided prior to the study. NIST reported values for Cr, Mo, and Se using microwave digestion and inductively coupled plasma mass spectrometry (ICP-MS) with standard additions as the method of quantitation. The NIST values in whole egg powder are reported in the table below with an estimated relative uncertainty of 5 %.

Analyte	Estimated Mass Fraction (mg/kg) (as-received basis)		
Cr	0.687	±	0.034
Mo	0.581	±	0.029
Se	1.40	±	0.07

Study Results

- Fifty-three laboratories enrolled in this exercise and received samples with a minimum of 37 laboratories reporting results for one or more elements (70 % participation).
- The consensus means for chromium and molybdenum in the multivitamin/multielement tablets were within the target range with an acceptable variability (14 % and 16 % relative standard deviation (RSD), respectively). The consensus mean for selenium in the multivitamin/multielement tablets was below the target range, with a slightly higher variability (19 % RSD).
- The consensus means for molybdenum and selenium in the whole egg powder were within the target range. While molybdenum had an acceptable variability (14 % RSD), the variability for selenium was higher (26 % RSD). The consensus mean for chromium in the whole egg powder was above the target range with an unacceptable variability of 63 % RSD.
- A majority of the laboratories reported using either open-beaker digestion (29 % to 36 %, depending on the element) or microwave digestion (52 % to 58 %) for sample preparation. The remaining laboratories reported using hot block digestion (11% to 13 %).
- A majority of the laboratories reported using either ICP-MS (72 % to 81 %, depending on the element) or ICP-OES (18 % to 23 %) as their analytical method. Less than 5 % of the laboratories reported using atomic absorption spectroscopy (AAS) or total reflection X-ray fluorescence (TXRF).

Technical Recommendations

The following recommendations are based on results reported by the participants in this study.

- There did not seem to be a difference in results based on either open-beaker digestions or microwave digestions for the elements in this study. There also did not appear to be any difference in results based on either ICP-OES or ICP-MS analytical methods. (Too few results were reported by other methods to identify any trends).
  - Laboratories that reported high values for one material and low values for the second material for any particular element (see **Figure 7**, **Figure 8**, and **Figure 9**) may have more trouble digesting one sample matrix over the other. SRM 3280 Multivitamin/Multielement Tablets are very difficult to digest, requiring relatively high temperatures, regardless of digestion method, to get complete sample dissolution. Laboratories using higher temperatures for digestions were more consistent at reporting values within consensus or target ranges for all elements.
  - It is important to note that with different sample matrices, there may also be different interferences to take into consideration during sample analysis.
- The elongated consensus box in **Figure 7** is due to several high values reported for the whole egg powder. There are several possibilities for this, one being calibration errors.

- The concentrations of these three elements in whole egg powder was approximately 10 to 100 times less than those in SRM 3280 Multivitamin/Multielement Tablets so there was the possibility of contamination if the two materials were prepared together.
- With both ICP-OES and ICP-MS, it is important to check the calibration curve for linearity within the range of the sample solutions.
- With ICP-OES, some elements will not be linear beyond an upper limit. Make sure solution concentrations fall within that linear range.
- With ICP-MS, many instruments run in pulse mode, which is more sensitive. If the calibration curve extends beyond the dynamic range for pulse mode then the instrument will use both the pulse and analog mode. The ICP-MS must be calibrated for both modes in this case. It is often easier and more accurate to have a narrower range of calibration points, making sure the calibration curve is linear in the pulse mode.
- Run a quality control sample of known accuracy to ensure your method is performing as expected.
- Double-check all calculations for any errors.

Table 1. Individual data table (NIST) for trace nutritional elements in foods and dietary supplements.

National Institute of Standards & Technology

Exercise I – October 2012 – Nutritional Elements

Lab Code:		NIST	1. Your Results				2. Community Results			3. Target	
Analyte	Sample	Units	x_i	s_i	Z_{comm}	Z_{NIST}	N	x^*	s^*	x_{NIST}	U_{95}
Cr	Multivitamin Tablet	μg/g	92.4	2.7	0.2	0.0	41	89.8	12.7	92.4	2.7
Cr	Egg Powder	μg/g	0.687	0.034	-0.2	0.0	38	0.808	0.511	0.687	0.034
Mo	Multivitamin Tablet	μg/g	69.7	4.4	-0.1	0.0	39	70.4	11.2	69.7	4.4
Mo	Egg Powder	μg/g	0.581	0.029	0.0	0.0	35	0.580	0.083	0.581	0.029
Se	Multivitamin Tablet	μg/g	17.2	0.4	0.4	0.0	39	16.1	3.0	17.2	0.4
Se	Egg Powder	μg/g	1.40	0.07	-0.1	0.0	36	1.44	0.38	1.40	0.07

x_i Mean of reported values

s_i Standard deviation of reported values

Z_{comm} Z-score with respect to community consensus

Z_{NIST} Z-score with respect to NIST value

N   Number of quantitative values reported

x^* Robust mean of reported values

s^* Robust standard deviation

x_{NIST} NIST-assessed value

U_{95} ±95% confidence interval about the assessed value or standard deviation (s_{NIST})

Table 2. Data summary table for chromium in foods and dietary supplements.

		Chromium									
		SRM 3280 Multivitamin Tablet (µg/g)					Whole Egg Powder (µg/g)				
	Lab	A	B	C	Avg	SD	A	B	C	Avg	SD
Individual Results	NIST				92 4	2 7				0 687	0 034
	1901	70 3	65 1	67 6	67 7	2 6	0 610	0 560	0 640	0 603	0 040
	1903	99 5	97 6	98 2	98 4	1 0	0 536	0 527	0 528	0 530	0 005
	1904										
	1906										
	1907										
	1908	92 5	89 2	88 7	90 1	2 1	0 451	0 444	0 454	0 450	0 005
	1910	92 4	101 0	97 9	97 1	4 4	1 060	1 190	1 220	1 157	0 085
	1911	89 9	90 4	93 7	91 3	2 1	0 450	0 410	0 440	0 433	0 021
	1915										
	1917	89 5	88 2	89 7	89 1	0 8	1 318	1 368	1 414	1 367	0 048
	1920	111 5	110 8	113 0	111 8	1 1	0 660	0 656	0 650	0 655	0 005
	1925	77 3	84 0	92 6	84 7	7 7	1 280	1 372	1 789	1 480	0 271
	1928	82 1	76 8	80 8	79 9	2 8	1 200	1 100	1 200	1 167	0 058
	1930	95 9	108 8	107 3	104 0	7 0	0 518	0 522	0 675	0 572	0 090
	1931	77 7	89 5	88 1	85 1	6 4	0 631	0 553	0 520	0 568	0 057
	1932	79 3	84 7	85 1	83 0	3 2	0 450	0 450	0 469	0 456	0 011
	1933	98 0	99 0	101 0	99 3	1 5					
	1934	89 5	84 0	92 7	88 7	4 4	8 873	9 132	3 049	7 018	3 439
	1935										
	1936	79 0	77 0	82 0	79 3	2 5	0 430	0 390	0 510	0 443	0 061
	1938										
	1939	72 8			72 8		0 801			0 801	
	1940	95 3	102 4	86 9	94 9	7 8	1 393	0 557	0 659	0 869	0 456
	1941	93 2	92 7	86 2	90 7	3 9	0 418	0 315	0 279	0 337	0 072
	1942	80 0	84 0	84 0	82 7	2 3	1 200	1 200	1 200	1 200	0 000
	1943	67 4	65 0	70 9	67 8	3 0	0 360	0 350	0 340	0 350	0 010
	1944										
	1947	87 4	108 9	112 9	103 0	13 7	0 432	0 463	0 474	0 456	0 022
	1948	81 4	86 6	91 9	86 6	5 3	0 400	0 400	0 400	0 400	0 000
	1949	84 8	84 9	80 5	83 4	2 5	1 800	2 000	2 600	2 133	0 416
	1950										
	1951	93 0	88 8	95 1	92 3	3 2	0 434	0 417	0 409	0 420	0 013
	1953	89 3	121 8	103 5	104 9	16 3	0 638	0 631	0 683	0 651	0 028
	1954	93 9	94 9	87 0	91 9	4 3					
	1955	95 9	98 2	95 1	96 4	1 6	2 621	3 323	3 612	3 185	0 510
	1956	87 7	92 1	94 8	91 5	3 6	2 917	3 646	2 896	3 153	0 427
	1958	147 9	151 3	149 2	149 5	1 7	0 200	0 200	0 200	0 200	0 000
	1959	91 0	93 0	93 8	92 6	1 5	0 569	0 515	0 601	0 562	0 044
	1960	92 9	103 9		98 4	7 8	0 461	0 461	0 425	0 449	0 021
	1961	84 1	84 3	80 0	82 8	2 4	0 807	0 775	0 745	0 776	0 031
	1963	98 1	92 2	91 9	94 1	3 5					
	1964	71 9	68 4	70 8	70 4	1 8	0 480	0 520	0 500	0 500	0 020
	1965										
	1966	55 9	56 4	48 6	53 7	4 4	1 600	1 040	1 040	1 227	0 323
	1967	101 0	105 0	108 0	104 7	3 5	1 770	1 500	1 800	1 690	0 165
	1972	97 1	98 4	98 2	97 9	0 7	0 472	0 505	0 489	0 489	0 017
	1973	79 0	81 7	90 2	83 6	5 8	0 465	0 432	0 431	0 443	0 019
	1978	62 3	70 3	69 5	67 4	4 4	0 518	0 484	0 589	0 530	0 054
	1980	108 0	95 0	97 9	100 3	6 8	0 795	0 683	0 730	0 736	0 056
	1981	82 9	76 3	79 8	79 6	3 3	2 280	2 650	2 280	2 403	0 214
	1983	108 0	105 0		106 5	2 1					
	1985	71 7			71 7		0 780			0 780	
	1986						0 673	0 714	0 674	0 687	0 023
Community Results		Consensus Mean			89 8		Consensus Mean			0 808	
		Consensus Standard Deviation			12 7		Consensus Standard Deviation			0 511	
		Maximum			149 5		Maximum			7 018	
		Minimum			53 7		Minimum			0 200	
		N			41		N			38	

Table 3. Data summary table for molybdenum in foods and dietary supplements.

	Lab	SRM 3280 Multivitamin Tablet (µg/g)					Whole Egg Powder (µg/g)				
		A	B	C	Avg	SD	A	B	C	Avg	SD
Individual Results	NIST				69 7	4 4				0 581	0 029
	1901										
	1903	71 3	68 7	66 3	68 8	2 5	0 615	0 598	0 608	0 607	0 009
	1904										
	1906										
	1907										
	1908	86 5	71 5	84 4	80 8	8 1	0 716	0 495	0 440	0 550	0 146
	1910	65 0	62 5	70 2	65 9	3 9	1 090	0 632	0 511	0 744	0 305
	1911	66 7	65 6	63 2	65 2	1 8	0 556	0 522	0 554	0 544	0 019
	1915										
	1917	72 8	72 3	70 6	71 9	1 1	0 581	0 582	0 591	0 585	0 006
	1920	76 7	62 1	69 3	69 3	7 3	0 606	0 622	0 600	0 609	0 011
	1925	64 9	69 9	74 5	69 8	4 8	1 386	1 352	1 756	1 498	0 224
	1928	89 1	81 7	83 8	84 9	3 8	0 500	0 500	0 500	0 500	0 000
	1930	55 4	58 2	56 1	56 6	1 4	0 531	0 523	0 492	0 515	0 021
	1931	78 7	67 6	75 6	74 0	5 7	0 578	0 570	0 575	0 574	0 004
	1932	94 2	108 4	108 7	103 8	8 3	0 640	0 640	0 633	0 638	0 004
	1933	70 0	74 0	72 0	72 0	2 0					
	1934	66 6	55 7	64 9	62 4	5 9	0 678	0 747	0 598	0 674	0 074
	1935										
	1936	86 0	73 0	74 0	77 7	7 2	0 570	0 550	0 560	0 560	0 010
	1938										
	1939	52 8			52 8		0 361			0 361	
	1940	74 4	110 1	70 5	85 0	21 8	0 774	0 604	0 540	0 639	0 121
	1941	61 7	61 9	62 6	62 1	0 5	0 186	0 159	0 146	0 164	0 021
	1942	61 0	65 0	68 0	64 7	3 5	0 550	0 550	0 530	0 543	0 012
	1943	65 5	65 2	67 8	66 2	1 4	0 520	0 520	0 510	0 517	0 006
	1944										
	1947	76 2	80 7	78 3	78 4	2 3	0 544	0 498	0 527	0 523	0 023
	1948	71 1	72 4	77 9	73 8	3 6	0 570	0 580	0 580	0 577	0 006
	1949	49 5	54 6	53 1	52 4	2 6	0 900	0 500	0 700	0 700	0 200
	1950										
	1951	87 5	81 0	70 4	79 6	8 6	0 695	0 658	0 686	0 680	0 019
	1953	88 0	78 8	71 4	79 4	8 3	0 566	0 581	0 585	0 577	0 010
	1954	60 0	67 7	60 7	62 8	4 3					
	1955	72 0	65 6	72 7	70 1	3 9	0 449	0 453	0 447	0 450	0 003
	1956	67 0	90 7	56 0	71 2	17 8	0 467	0 472	0 459	0 466	0 007
	1958	87 5	84 8	82 2	84 8	2 7					
	1959	61 5	55 7	55 9	57 7	3 3	0 606	0 579	0 607	0 597	0 016
	1960	64 8	87 2		76 0	15 8	0 514	0 548	0 488	0 517	0 030
	1961	63 0	62 2	15 0	46 7	27 5	0 798	0 776	0 366	0 647	0 244
	1963	76 3	79 8	76 7	77 6	1 9					
	1964	84 4	86 3	85 8	85 5	1 0	0 600	0 620	0 600	0 607	0 012
	1965										
	1966										
	1967	70 4	58 5	79 9	69 6	10 7	0 630	0 530	0 630	0 597	0 058
	1972	67 5	60 4	71 4	66 4	5 6	0 557	0 559	0 549	0 555	0 005
	1973	72 3	82 1	75 1	76 5	5 0	0 515	0 494	0 507	0 505	0 011
	1978	47 8	52 3	52 9	51 0	2 8	0 622	0 541	0 698	0 620	0 078
	1980	77 0	72 9	80 2	76 7	3 7	0 602	0 613	0 602	0 606	0 006
	1981	82 8	70 3	77 8	77 0	6 2	0 820	0 640	0 590	0 683	0 121
	1983	53 0	56 0		54 5	2 1					
	1985	50 1			50 1		0 460			0 460	
	1986						0 592	0 581	0 571	0 581	0 011
Community Results		Consensus Mean			70 4		Consensus Mean			0 580	
		Consensus Standard Deviation			11 2		Consensus Standard Deviation			0 083	
		Maximum			104		Maximum			1 498	
		Minimum			47		Minimum			0 164	
		N			39		N			35	

Table 4. Data summary table for selenium in foods and dietary supplements.

	Lab	SRM 3280 Multivitamin Tablet (µg/g)					Whole Egg Powder (µg/g)				
		A	B	C	Avg	SD	A	B	C	Avg	SD
Individual Results	NIST				17 2	0 4				1 40	0 07
	I901										
	I903	16 4	15 7	17 0	16 4	0 7	1 39	1 40	1 42	1 40	0 02
	I904										
	I906										
	I907										
	I908	18 4	17 9	16 0	17 4	1 3	1 65	1 59	1 57	1 60	0 04
	I910	17 7	18 0	16 2	17 3	1 0	0 93	1 50	1 13	1 19	0 29
	I911	17 0	14 8	15 6	15 8	1 1	1 60	1 50	1 50	1 53	0 06
	I915										
	I917	14 9	15 1	14 6	14 9	0 2	2 14	2 20	2 06	2 13	0 07
	I920	12 7	11 6	11 2	11 8	0 8	1 45	1 49	1 44	1 46	0 03
	I925	17 8	17 7	17 5	17 7	0 1	1 40	1 37	1 46	1 41	0 05
	I928	24 9	20 6	22 7	22 7	2 2	1 80	1 90	1 90	1 87	0 06
	I930	14 0	14 9	13 8	14 2	0 6	1 67	1 53	1 49	1 56	0 09
	I931	16 2	15 4	15 7	15 7	0 4	1 74	1 57	1 68	1 66	0 09
	I932	11 7	12 2	13 6	12 5	1 0	1 58	1 60	1 48	1 55	0 07
	I933										
	I934	18 7	17 5	18 1	18 1	0 6	1 43	0 95	0 94	1 11	0 28
	I935										
	I936	20 0	18 0	19 0	19 0	1 0	1 50	1 70	1 60	1 60	0 10
	I938										
	I939	8 8			8 8		1 81			1 81	
	I940	11 0	13 2	12 2	12 1	1 1	0 97	0 09	0 97	0 68	0 51
	I941	14 9	15 1	15 2	15 0	0 1	0 73	0 76	0 76	0 75	0 02
	I942	15 0	16 0	17 0	16 0	1 0	1 70	1 70	1 60	1 67	0 06
	I943	18 6	15 9	16 4	17 0	1 4	1 36	1 40	1 20	1 32	0 11
	I944										
	I947	23 1	22 5	20 7	22 1	1 3	1 47	1 37	1 49	1 44	0 06
	I948	16 3	15 5	17 8	16 5	1 2	1 40	1 40	1 40	1 40	0 00
	I949	15 3	15 3	14 5	15 0	0 5	1 20	1 00	1 30	1 17	0 15
	I950										
	I951	19 3	15 6	17 5	17 4	1 9	1 44	1 35	1 39	1 39	0 05
	I953	6 5	9 3	6 4	7 4	1 6	0 83	0 83	0 82	0 83	0 01
	I954	16 2	17 1	15 4	16 2	0 9					
	I955	17 8	16 3	16 8	17 0	0 8	1 69	1 71	1 70	1 70	0 01
	I956	19 5	17 1	17 4	18 0	1 3	1 76	1 82	1 74	1 77	0 04
	I958	19 9	19 8	21 1	20 3	0 7	0 90	1 00	1 00	0 97	0 06
	I959	13 1	10 3	11 2	11 5	1 4					
	I960	21 1	19 9		20 5	0 9	1 39	1 44	1 32	1 39	0 06
	I961	16 0	15 0	15 4	15 4	0 5					
	I963	18 8	18 4	17 4	18 2	0 7	2 21	2 23	2 19	2 21	0 02
	I964	13 0	15 4	13 2	13 9	1 3					
	I965										
	I966	21 3	22 1	19 0	20 8	1 6	0 80	0 48	1 04	0 77	0 28
	I967	15 6	20 1	17 1	17 6	2 3	1 86	1 62	1 86	1 78	0 14
	I972	19 5	14 5	17 0	17 0	2 5	1 30	1 27	1 43	1 33	0 08
	I973	15 7	14 9	15 1	15 2	0 4	1 23	1 31	1 37	1 30	0 07
	I978	2 8	2 3	3 2	2 8	0 5	1 00	1 08	1 01	1 03	0 05
	I980	14 4	12 4	12 9	13 2	1 0	2 04	1 99	1 98	2 00	0 03
	I981	16 9	16 0	14 4	15 8	1 3	1 37	1 41	1 43	1 40	0 03
	I983	17 5	16 0	16 0	16 5	0 9	1 40	1 50	1 50	1 47	0 06
	I985	9 4			9 4		2 00			2 00	
	I986						1 43	1 41	1 37	1 40	0 03
Community Results		Consensus Mean			16 1		Consensus Mean			1 44	
		Consensus Standard Deviation			3 0		Consensus Standard Deviation			0 38	
		Maximum			22 7		Maximum			2 21	
		Minimum			2 8		Minimum			0 68	
		N			39		N			36	

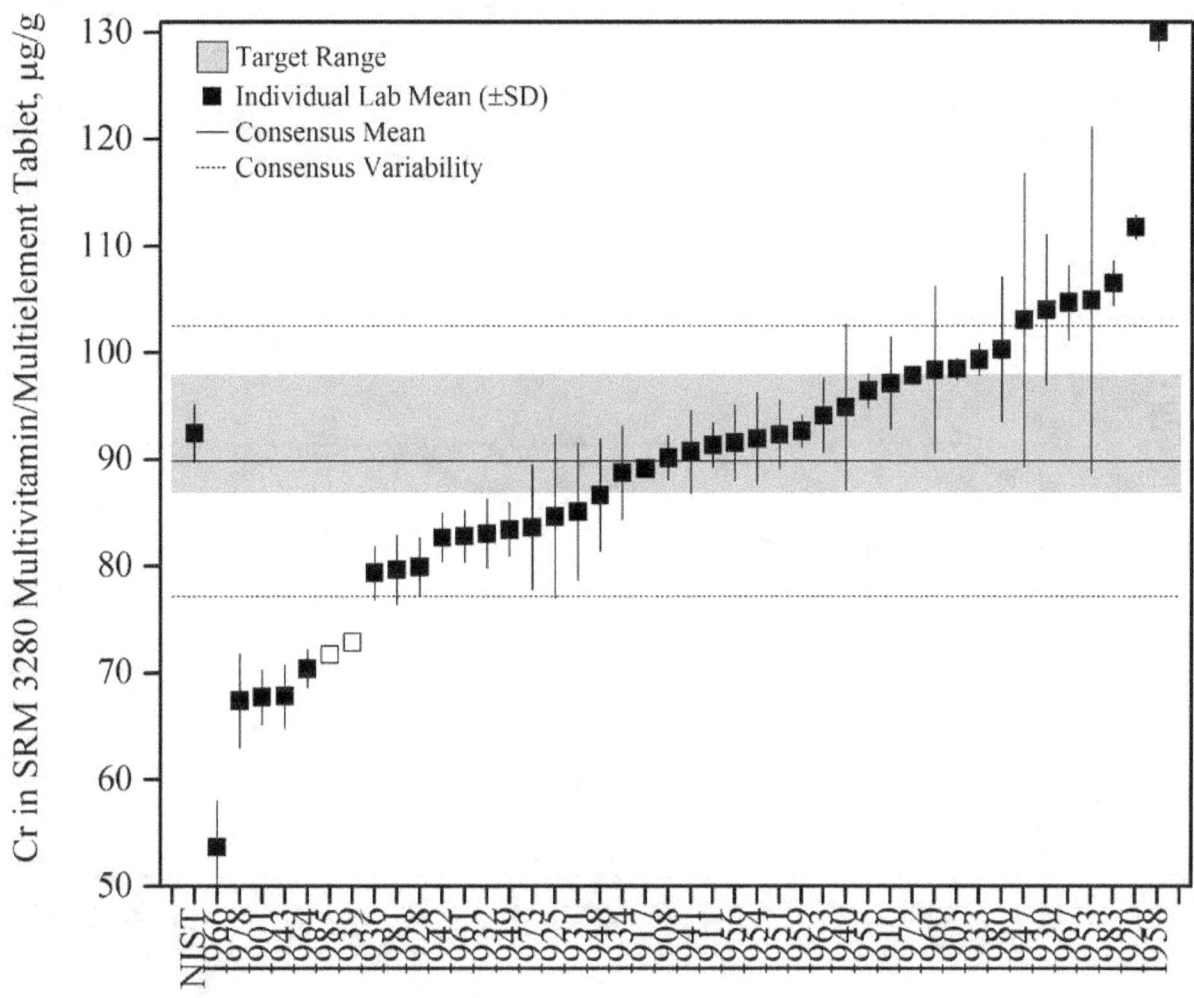

Figure 1. Chromium in SRM 3280 Multivitamin/Multielement Tablets (method comparison data summary view – digestion method). In this view, individual laboratory data are plotted with the individual laboratory standard deviation (error bars). Data points that are unfilled represent laboratories that only reported a single value for that analyte and therefore were not included in the consensus mean. The black solid line represents the consensus mean, and the black dotted lines represent the consensus variability calculated as one standard deviation about the consensus mean. The gray shaded region represents the target zone for "acceptable" performance, which encompasses the NIST certified value bounded by twice its uncertainty (U_{95}).

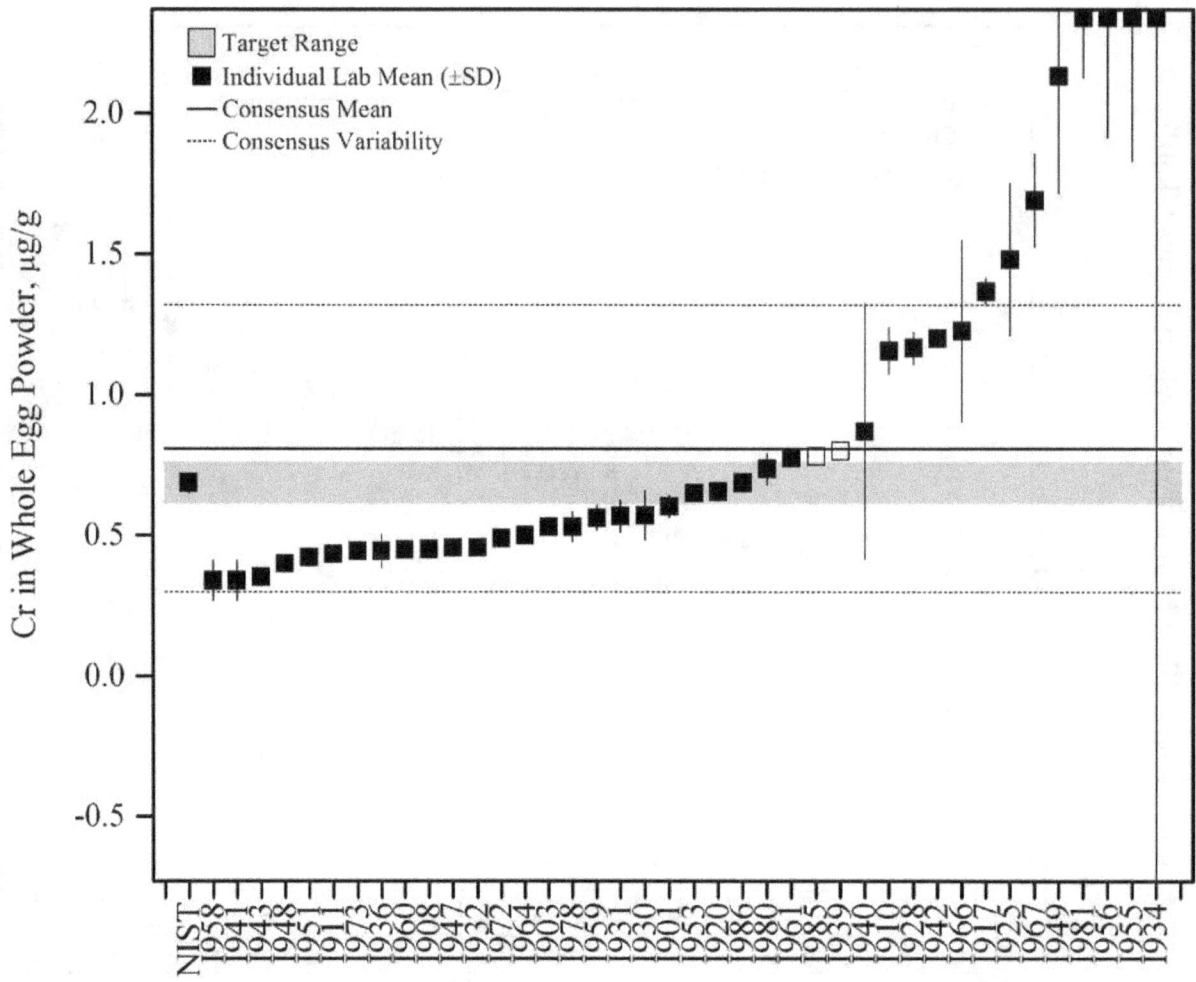

Figure 2. Chromium in whole egg powder (data summary view). In this view, individual laboratory data are plotted with the individual laboratory standard deviation (error bars). Data points that are unfilled represent laboratories that only reported a single value for that analyte and therefore were not included in the consensus mean. The black solid line represents the consensus mean, and the black dotted lines represent the consensus variability calculated as one standard deviation about the consensus mean. The gray shaded region represents the target zone for "acceptable" performance, which encompasses the NIST-assessed value bounded by an uncertainty of 5 %. The NIST value is the mean of three results determined by ICP-MS.

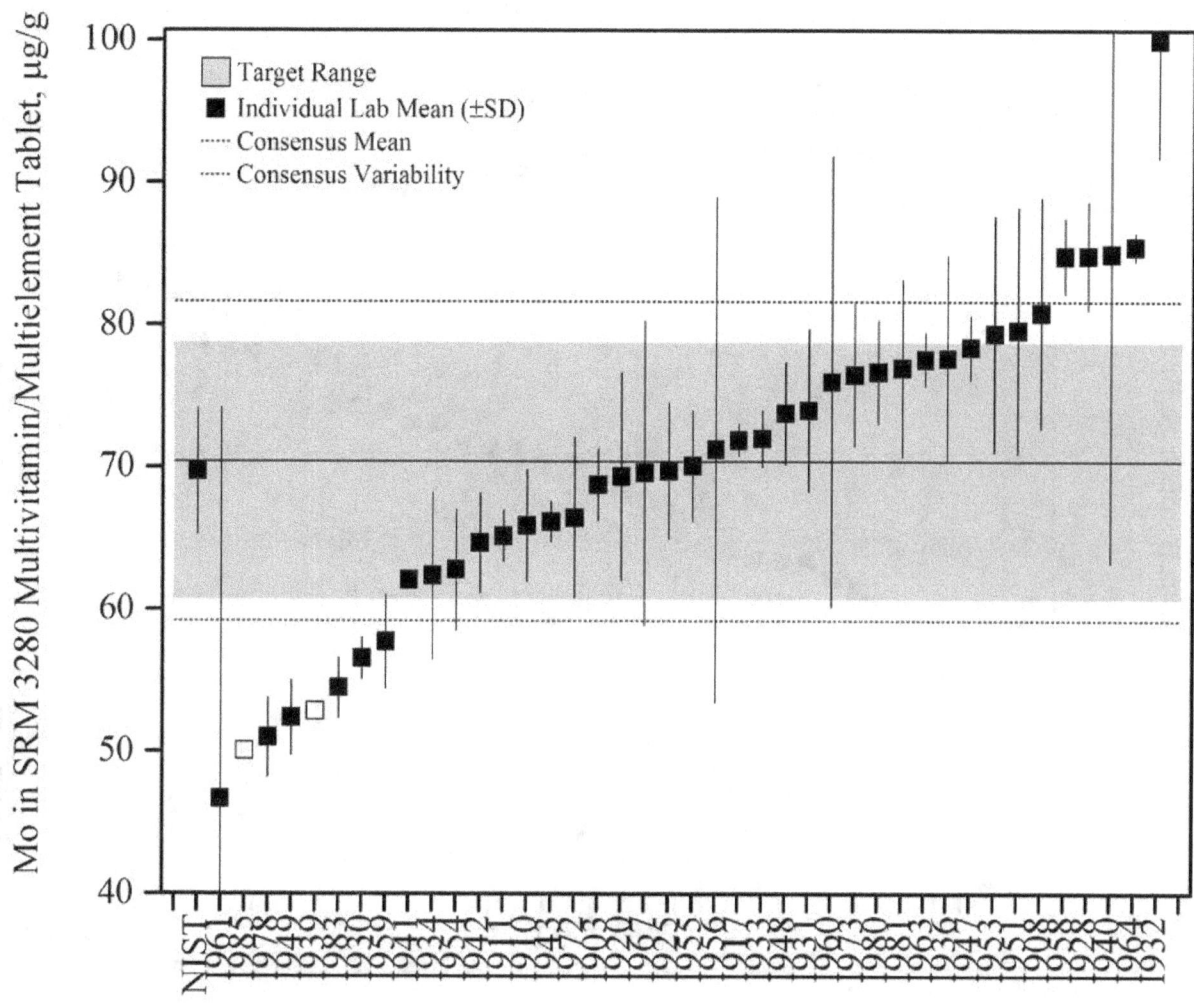

Figure 3. Molybdenum in SRM 3280 Multivitamin/Multielement Tablets (data summary view). In this view, individual laboratory data are plotted with the individual laboratory standard deviation (error bars). Data points that are unfilled represent laboratories that only reported a single value for that analyte and therefore were not included in the consensus mean. The black solid line represents the consensus mean, and the black dotted lines represent the consensus variability calculated as one standard deviation about the consensus mean. The gray shaded region represents the target zone for "acceptable" performance, which encompasses the NIST certified value bounded by twice its uncertainty (U_{95}).

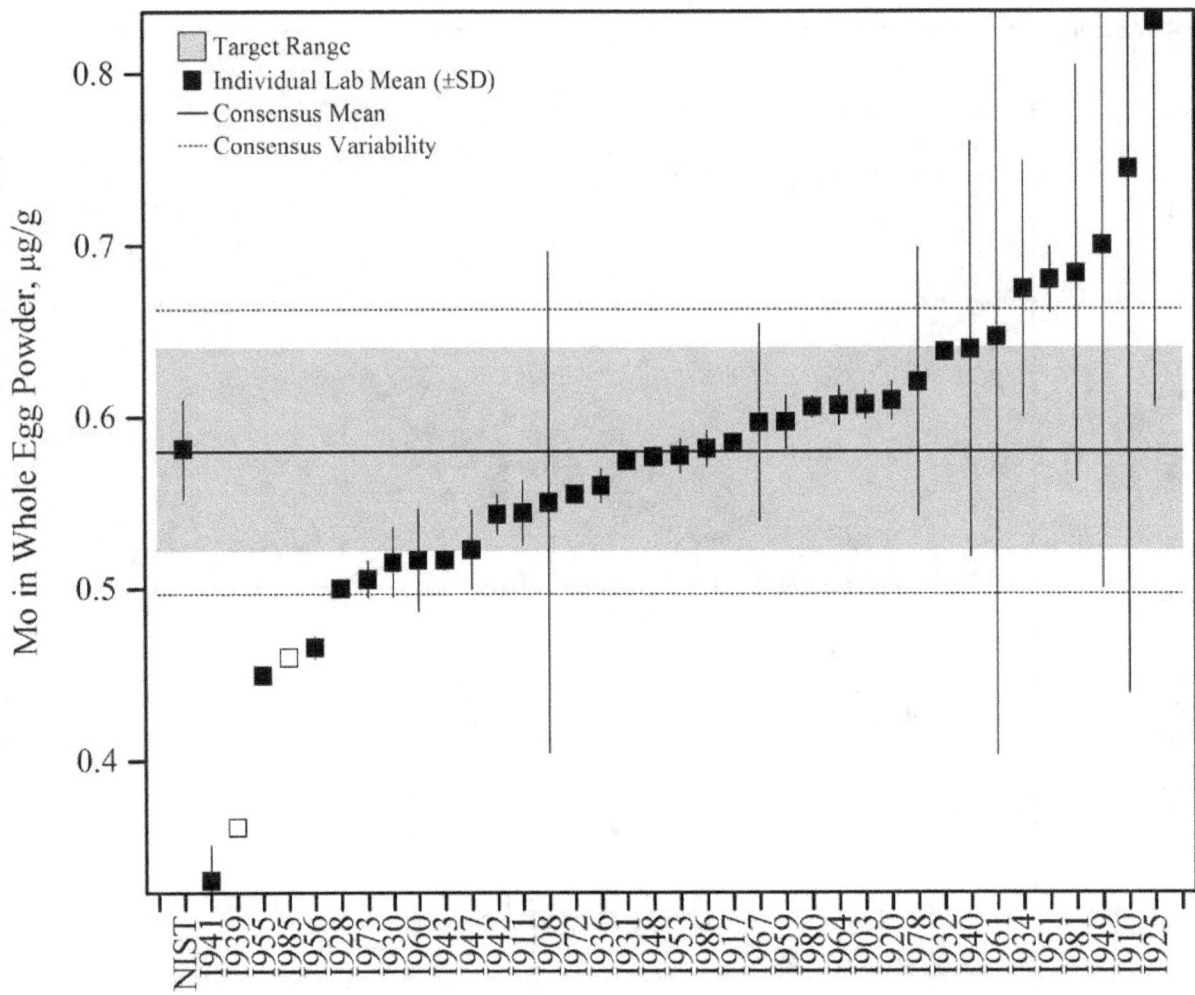

Figure 4. Molybdenum in whole egg powder (data summary view). In this view, individual laboratory data are plotted with the individual laboratory standard deviation (error bars). Data points that are unfilled represent laboratories that only reported a single value for that analyte and therefore were not included in the consensus mean. The black solid line represents the consensus mean, and the black dotted lines represent the consensus variability calculated as one standard deviation about the consensus mean. The gray shaded region represents the target zone for "acceptable" performance, which encompasses the NIST-assessed value bounded by an uncertainty of 5 %. The NIST value is the mean three of results determined by ICP-MS.

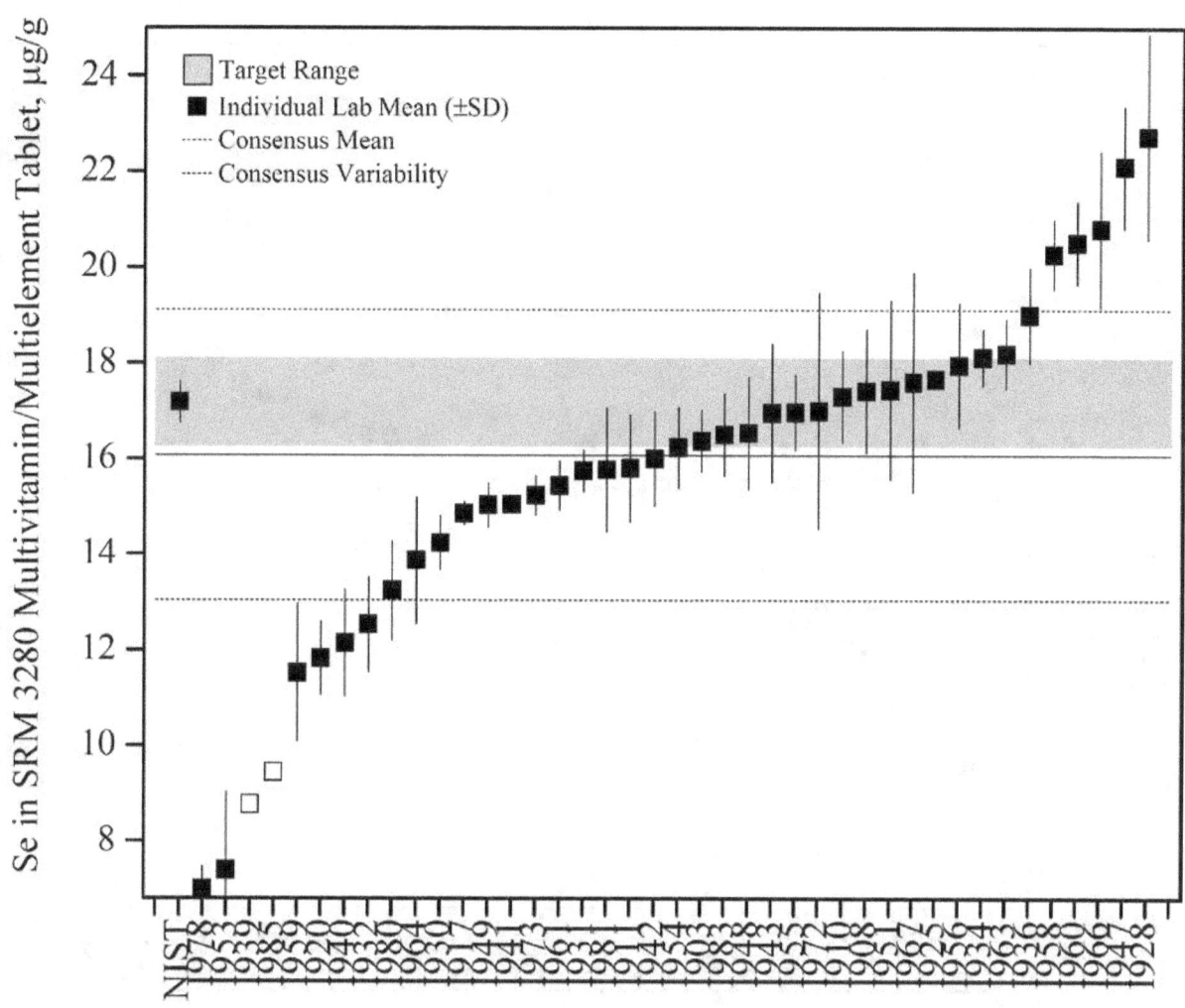

Figure 5. Selenium in SRM 3280 Multivitamin/Multielement Tablets (data summary view). In this view, individual laboratory data are plotted with the individual laboratory standard deviation (error bars). Data points that are unfilled represent laboratories that only reported a single value for that analyte and therefore were not included in the consensus mean. The black solid line represents the consensus mean, and the black dotted lines represent the consensus variability calculated as one standard deviation about the consensus mean. The gray shaded region represents the target zone for "acceptable" performance, which encompasses the NIST certified value bounded by twice its uncertainty (U_{95}).

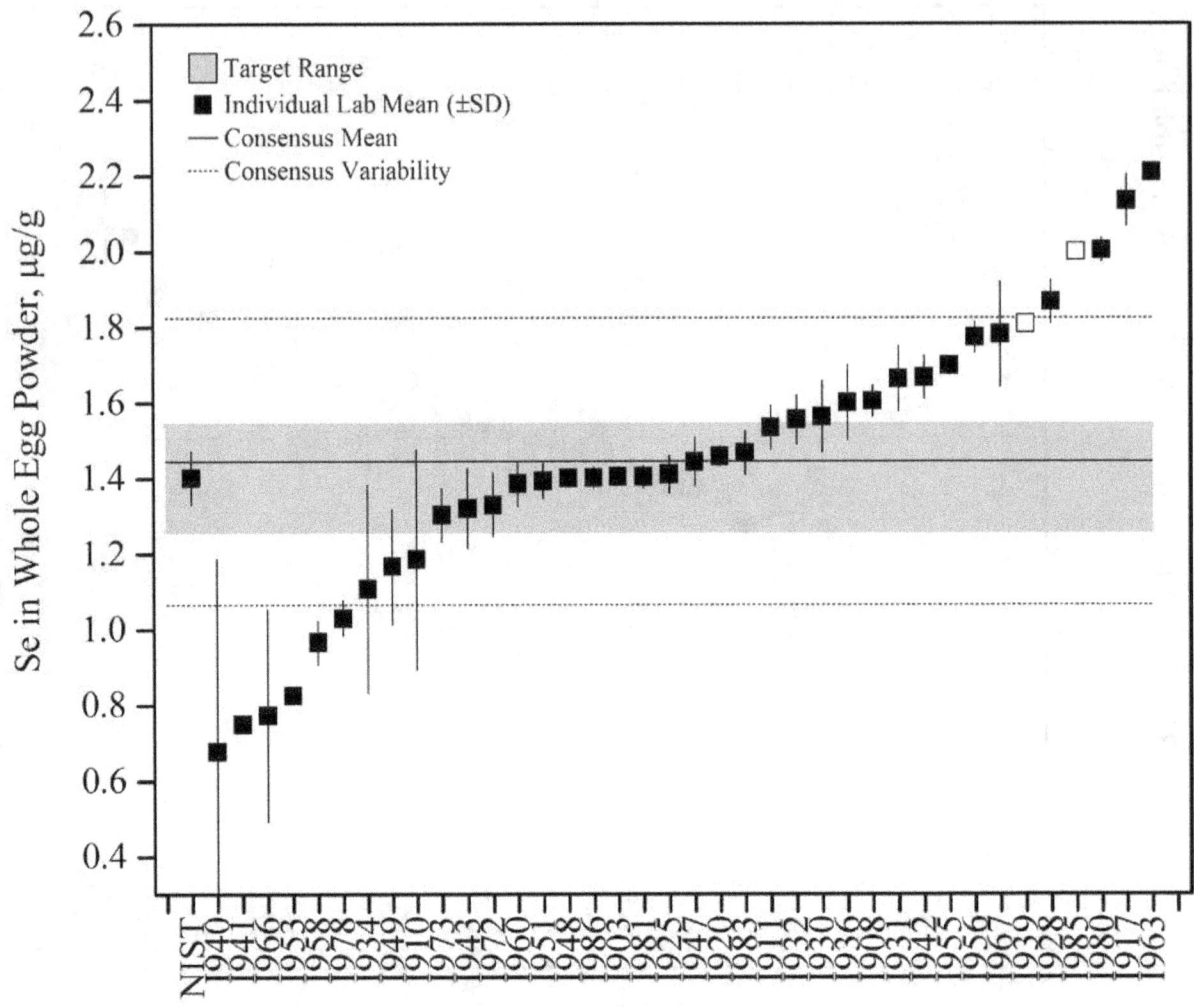

Figure 6. Selenium in whole egg powder (data summary view). In this view, individual laboratory data are plotted with the individual laboratory standard deviation (error bars). Data points that are unfilled represent laboratories that only reported a single value for that analyte and therefore were not included in the consensus mean. The black solid line represents the consensus mean, and the black dotted lines represent the consensus variability calculated as one standard deviation about the consensus mean. The gray shaded region represents the target zone for "acceptable" performance, which encompasses the NIST-assessed value bounded by an uncertainty of 5 %. The NIST value is the mean of three results determined by ICP-MS.

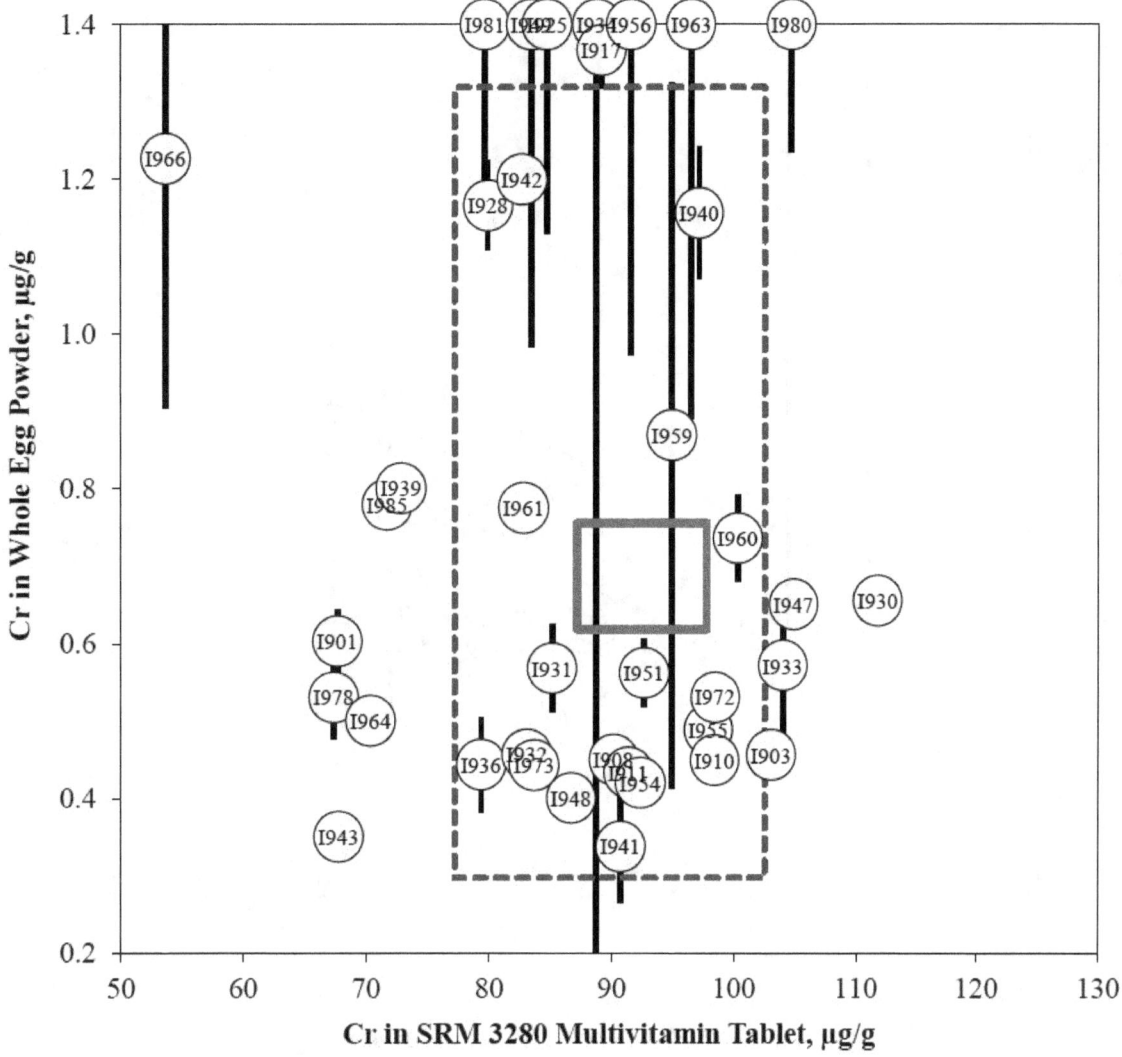

Figure 7. Chromium in whole egg powder and SRM 3280 Multivitamin/Multielement Tablets (sample/control comparison view). In this view, the individual laboratory results for the control (SRM 3280 Multivitamin/Multielement Tablets) with a certified value for the analyte are compared to the results for a sample (whole egg powder). The error bars represent the individual laboratory standard deviation. The solid red lines represent the target zone for the control (x-axis) and the unknown sample (y-axis). The dotted blue box represents the consensus zone for the control (x-axis) and the unknown sample (y-axis).

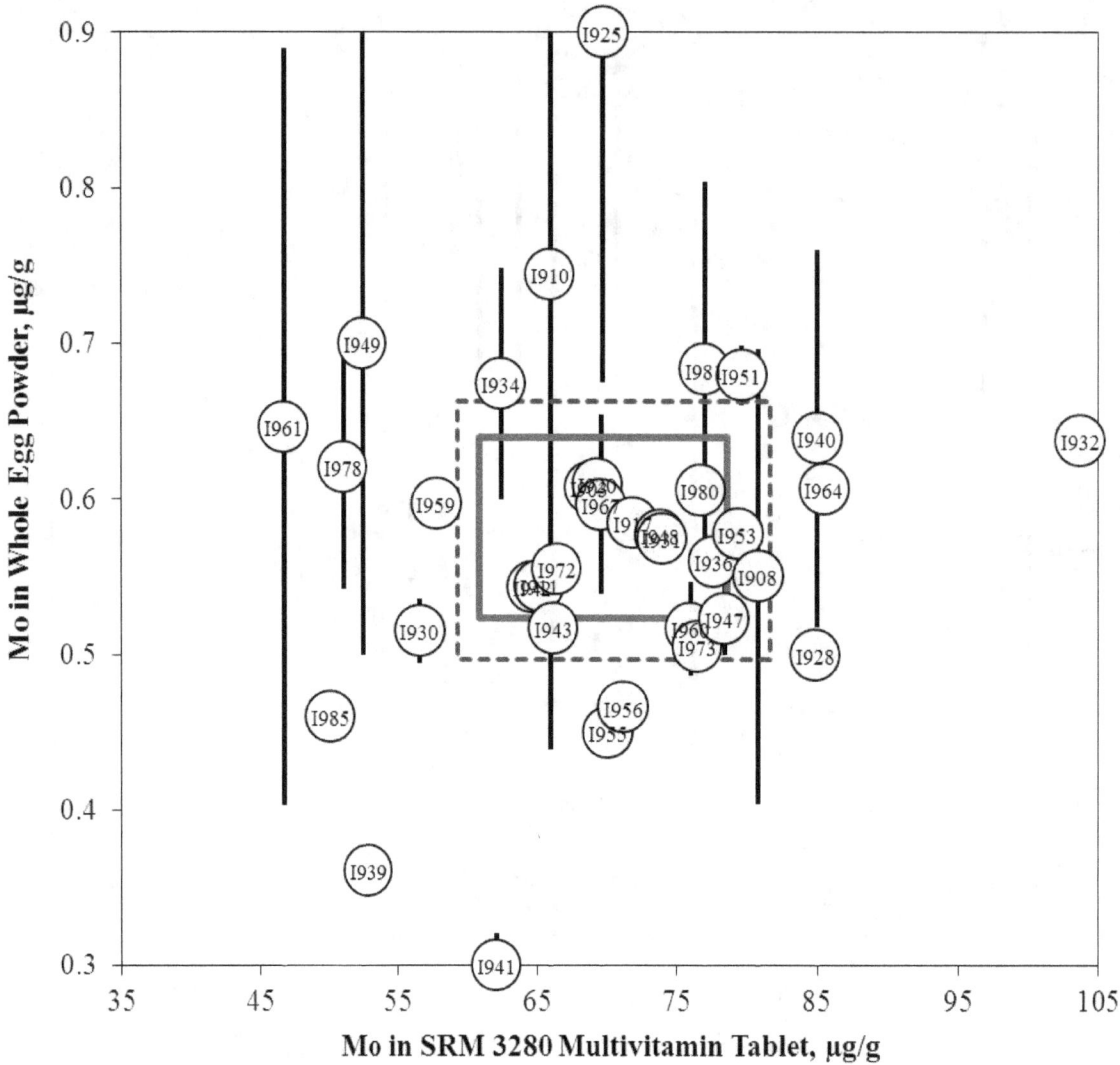

Figure 8. Molybdenum in whole egg powder and SRM 3280 Multivitamin/Multielement Tablets (sample/control comparison view). In this view, the individual laboratory results for the control (SRM 3280 Multivitamin/Multielement Tablets) with a certified value for the analyte are compared to the results for a sample (whole egg powder). The error bars represent the individual laboratory standard deviation. The solid red lines represent the target zone for the control (x-axis) and the unknown sample (y-axis). The dotted blue box represents the consensus zone for the control (x-axis) and the unknown sample (y-axis).

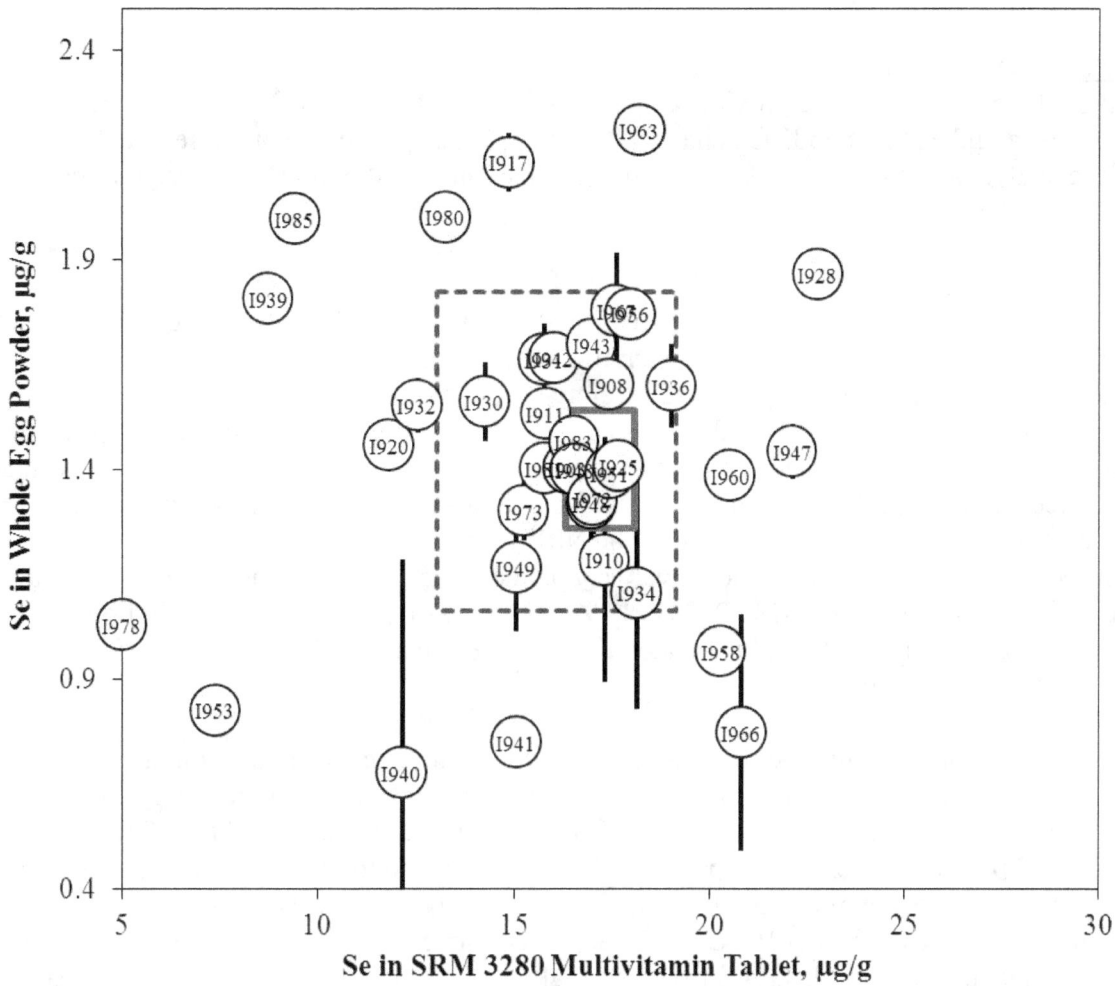

Figure 9. Selenium in whole egg powder and SRM 3280 Multivitamin/Multielement Tablets (sample/control comparison view). In this view, the individual laboratory results for the control (SRM 3280 Multivitamin/Multielement Tablets) with a certified value for the analyte are compared to the results for a sample (whole egg powder). The error bars represent the individual laboratory standard deviation. The solid red lines represent the target zone for the control (x-axis) and the unknown sample (y-axis). The dotted blue box represents the consensus zone for the control (x-axis) and the unknown sample (y-axis).

TOXIC ELEMENTS (Cd) IN FOODS AND SUPPLEMENTS

Study Overview

In this study, participants were provided with two NIST SRMs, SRM 3233 Fortified Breakfast Cereal and candidate SRM 3532 Calcium Dietary Supplement. Participants were asked to use in-house analytical methods to determine the mass fraction of cadmium (Cd) in each of the matrices and report values on an as-received basis.

Sample Information

Fortified breakfast cereal. Participants were provided with one packet containing approximately 10 g of fortified breakfast cereal. This material is a wheat-based fortified flake cereal that was ground to 180 μm, blended, and packaged. Before use, participants were instructed to mix the contents of the packet thoroughly and use a sample size of at least 0.5 g. Participants were asked to store the material at controlled room temperature, 10 °C to 30 °C, prepare three samples, and report three values from the single packet provided. Approximate analyte levels were not provided to participants prior to the study. The NIST certified value in SRM 3233 was determined using isotope dilution inductively coupled plasma mass spectrometry (ID-ICP-MS). The certified value for Cd in SRM 3233 is (81.9 ± 2.0) ng/g on a dry-mass basis. Following adjustment for moisture content of the material of 1.70 %, the as-received target value for Cd in SRM 3233 is (80.5 ± 2.0) ng/g.

Calcium dietary supplement. Participants were provided with one packet containing approximately 10 g of a powdered calcium dietary supplement. The calcium dietary supplement was prepared from commercially purchased calcium tablets that were ground to 180 μm, blended, and packaged. Before use, participants were instructed to thoroughly mix the contents of the packet and use a sample size of at least 0.5 g. Participants were asked to store the material at controlled room temperature, 10 °C to 30 °C, prepare three samples, and report three values from the single packet provided. Approximate analyte levels were not provided to participants prior to the study. The NIST-estimated value for Cd in candidate SRM 3532 was determined by ID-ICP-MS. The estimated value, based on the mean and expanded uncertainty of duplicate measurements from six packets, is (94.7 ± 1.7) ng/g.

Study Results
- Fifty-three laboratories enrolled in this exercise and received samples, and forty-two laboratories reported results for Cd (79 % participation).
- The consensus mean for Cd in the fortified breakfast cereal was within the target range with an acceptable variability (11 % RSD).
- The consensus mean for Cd in the calcium dietary supplement was slightly below the target range but had acceptable variability (16 % RSD).
- A majority of the laboratories reported using either microwave digestion (62 %) or open beaker digestion (29 %) for sample preparation. Four laboratories reported using hot block digestion (9 %).
- A majority of the laboratories (88 %) reported using ICP-MS as their analytical method for analysis. Only four laboratories reported using either ICP-OES or AAS for their analytical measurements for fortified breakfast cereal and five laboratories reported to

have used either ICP-OES or AAS for their analytical measurements for calcium dietary supplement (< 12 %).

Technical Recommendations

The following recommendations are based on results provided by the participants in this study.

- While twice as many laboratories reported using microwave digestion for sample preparation than other methods reported, there did not seem to be a difference in results based on the sample preparation method used.
- Cadmium can be difficult to measure by ICP-OES because of low sensitivity. Using AAS to measure Cd should not pose any significant problems but sometimes an extraction or separation step is included.
- Spectral interferences can make Cd difficult to measure by ICP-MS if there are high concentrations of certain elements, mainly Mo, Sn, or Zr, but the calcium dietary supplement presents the special case of having a high ratio of Ca to Cd [4].
  - A scan of the sample beforehand will indicate if there are potential interferences in the sample that will need to be addressed.
  - There can be interferences with commonly used masses of Cd (^{111}Cd, ^{112}Cd, ^{113}Cd, and ^{114}Cd). Examples of molecular interferences include: $^{95, 96, 97 \text{ and}}$ $^{98}Mo^{16}O^+$, $^{94, 95, 96, \text{ and } 97}Mo^{16}O^1H^+$, $^{96}Zr^{16}O^+$, $^{94 \text{ and } 96}Zr^{16}O^1H^+$, $^{40}Ar_2^{16}O_2$, $^{40}Ca_2^{16}O_2$, or $^{40}Ca_2^{16}O_2^1H^+$; examples of elemental isobaric interferences include: ^{112}Sn, ^{113}In, and ^{114}Sn.
  - Chemical separations by anion chromatography can reduce interferences but because of the labor intensive work involved it is usually impractical for laboratories to do a chemical separation on each sample.
  - Collision cell technology, available on most newer-model ICP-MS instruments, can be used to remove many of the molecular interferences that may be found in these two materials.
  - Interference equations inherent to the software provided on some ICP-MS instruments are designed to correct for interferences, and these equations can also be applied off-line. Both are less labor-intensive alternatives to chemical separations.
- Many ICP-MS instruments run in either pulse mode or analog mode.
  - If sample solutions fall outside of the dynamic range for pulse mode, then the instrument will use both the pulse and analog mode. In this case, the ICP-MS must be calibrated for both modes.
  - It is often easier and more accurate to ensure that the calibrants are linear in the pulse mode and that the samples are within this linear range.
  - As shown in **Figure 14**, many laboratories reported either high values for both samples or low values for both samples. High values may indicate spectral interference or contamination. Low values may indicate matrix-induced signal suppression which may be avoided with the use of an internal standard. Dilution of sample solutions can also decrease matrix-induced signal suppression as long as solutions are not diluted below the detection limit. Additionally, high or low

---

[4] Murphy, K.E., Vetter, T.W. (2013) *Recognizing and overcoming analytical error in the use of ICP-MS for the determination of cadmium in breakfast cereal and dietary supplements.* Anal Bioanal Chem **405** 4579-4588.

results can be an indication of a calibration error. A calibration curve needs to tightly bracket expected working solutions and be linear in that region. More accurate measurements can be achieved by making sure the sample concentrations fall within the middle of the calibration curve.

- Run a well-documented quality control sample with your unknown samples to ensure your method is performing as expected.
- Double-check all calculations for errors. Compare these to your quality assurance samples to make sure all calculations have been done correctly.

Table 5. Individual data table (NIST) for cadmium in foods and dietary supplements.

National Institute of Standards & Technology

Exercise I – October 2012 – Cd

Lab Code:		NIST	1. Your Results				2. Community Results			3. Target	
Analyte	Sample	Units	x_i	s_i	Z_{comm}	Z_{NIST}	N	x^*	s^*	x_{NIST}	U_{95}
Cd	Breakfast Cereal	ng/g	80.5	2.0	-0.7	0.0	39	80.4	9.0	80.5	2.0
Cd	Ca Supplement	ng/g	94.7	1.7	0.3	0.0	40	90.8	14.8	94.7	1.7

x_i Mean of reported values

s_i Standard deviation of reported values

Z_{comm} Z-score with respect to community consensus

Z_{NIST} Z-score with respect to NIST value

N   Number of quantitative values reported

x^* Robust mean of reported values

s^* Robust standard deviation

x_{NIST} NIST-assessed value

U_{95} ±95% confidence interval about the assessed value or standard deviation (s_{NIST})

Table 6. Data summary table for cadmium in foods and dietary supplements.

		Cadmium									
		SRM 3233 Fortified Breakfast Cereal (ng/g)					Candidate SRM 3532 Ca Supplement (ng/g)				
	Lab	A	B	C	Avg	SD	A	B	C	Avg	SD
Individual Results	NIST				80 5	2 0				94 7	1 7
	I901										
	I903	80 4	71 7	77 8	76 6	4 5	83 8	78 2	79 7	80 6	2 9
	I904										
	I906										
	I907										
	I908	72 0	70 0	78 0	73 3	4 2	79 5	81 5	84 5	81 8	2 5
	I910	65 0	55 5	61 9	60 8	4 8	117 0	104 0	122 0	114 3	9 3
	I911	88 0	98 0	86 0	90 7	6 4	98 0	85 0	83 0	88 7	8 1
	I915										
	I917	75 7	79 2	76 7	77 2	1 8	86 9	86 9	86 9	86 9	0 0
	I920	108 0	96 0	100 0	101 3	6 1	107 0	114 0	110 0	110 3	3 5
	I925	78 4	80 8	79 9	79 7	1 2	95 1	94 5	93 1	94 2	1 0
	I928	65 0	70 0	70 0	68 3	2 9	72 0	66 0	72 0	70 0	3 5
	I929	144 0	145 0	146 0	145 0	1 0	128 0	129 0	130 0	129 0	1 0
	I930	72 2	76 3	74 4	74 3	2 1	87 9	81 0	81 1	83 3	4 0
	I931	100 0	107 0	111 0	106 0	5 6	66 5	57 4	59 5	61 1	4 8
	I932	81 3	77 1	82 8	80 4	3 0	99 2	96 7	98 7	98 2	1 4
	I933	74 0	74 0	72 0	73 3	1 2	84 0	82 0	84 0	83 3	1 2
	I934	126 0	97 4	115 6	113 0	14 4	119 2	139 2	103 5	120 6	17 9
	I935										
	I936	80 0	85 0	85 0	83 3	2 9	91 0	100 0	100 0	97 0	5 2
	I938										
	I939	52 0			52 0		58 0			58 0	
	I940	76 7	84 3	73 2	78 1	5 6	63 1	69 3	99 0	77 1	19 1
	I942	77 0	81 0	88 0	82 0	5 6	75 0	74 0	76 0	75 0	1 0
	I943	84 0	79 0	81 0	81 3	2 5	96 0	92 0	91 0	93 0	2 6
	I944										
	I945	83 7	84 6	84 4	84 2	0 5	98 1	98 3	94 2	96 8	2 3
	I947	89 9	86 3	83 5	86 5	3 2	96 7	99 2	92 6	96 1	3 3
	I948	100 0	90 0	90 0	93 3	5 8	110 0	100 0	110 0	106 7	5 8
	I949										
	I950	58 7	58 1	57 6	58 1	0 6	61 5	62 7	63 6	62 6	1 1
	I951	83 3	90 8	85 3	86 5	3 9	95 6	93 9	100 1	96 5	3 2
	I953	72 0	78 0	74 0	74 7	3 1	91 0	91 0	92 0	91 3	0 6
	I954	80 0	84 0	80 0	81 3	2 3	90 0	97 0	91 0	92 7	3 8
	I955	73 5	72 4	75 4	73 8	1 5	78 2	81 0	80 5	79 9	1 5
	I956	75 2	76 3	76 4	75 9	0 7	78 3	82 0	81 1	80 5	2 0
	I958	74 0	80 0	78 0	77 3	3 1	110 0	99 0	124 0	111 0	12 5
	I959	80 6	77 7	76 7	78 3	2 0	100 7	95 1	94 1	96 6	3 6
	I960	83 3	77 5	85 4	82 1	4 1	86 0	102 3	90 7	93 0	8 4
	I961	162 2	127 2	112 4	134 0	25 6	254 9	204 7	211 6	223 7	27 2
	I962						74 6	69 7	77 6	74 0	4 0
	I963	85 4	85 6	81 4	84 1	2 4	101 3	93 7	96 4	97 2	3 8
	I964	72 0	75 0	71 0	72 7	2 1	81 0	81 0	84 0	82 0	1 7
	I966										
	I967	81 4	78 0	78 2	79 2	1 9	98 0	97 8	97 1	97 6	0 5
	I972	49 3	48 5	48 6	48 8	0 4	60 2	61 1	73 0	64 8	7 2
	I973	77 7	68 8	71 5	72 7	4 6	87 8	80 0	89 3	85 7	5 0
	I978	86 0	85 0	87 1	86 0	1 1	94 5	90 0	81 4	88 6	6 7
	I979										
	I980	80 0	79 0	77 0	78 7	1 5	98 0	90 0	92 0	93 3	4 2
	I983	73 0	79 0	89 0	80 3	8 1	95 0	92 0	98 0	95 0	3 0
	I985	57 0			57 0		62 0			62 0	
Community Results		Consensus Mean			80 4		Consensus Mean			90 8	
		Consensus Standard Deviation			9 0		Consensus Standard Deviation			14 8	
		Maximum			145 0		Maximum			223 7	
		Minimum			48 8		Minimum			58 0	
		N			39		N			40	

26

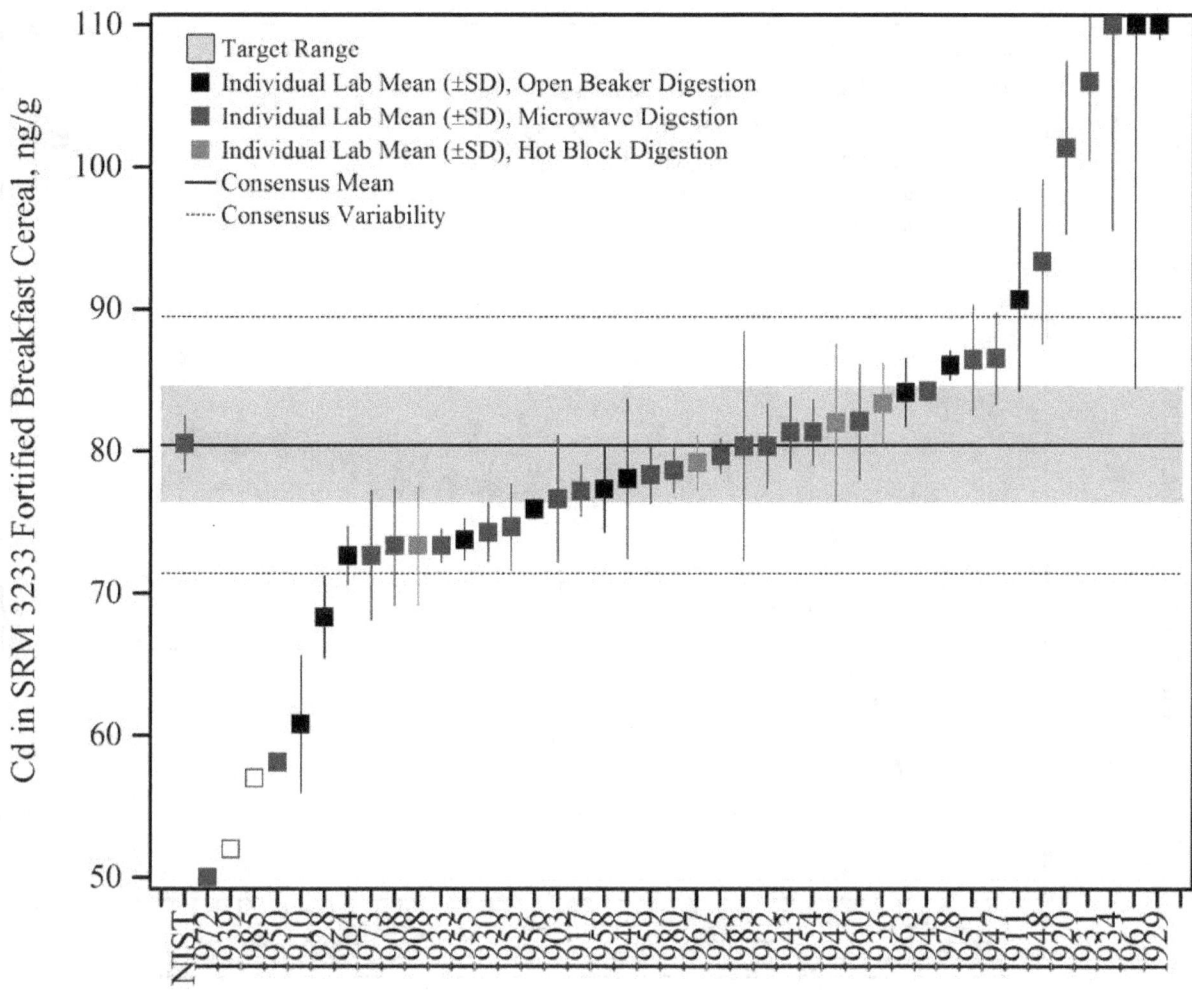

Figure 10. Cadmium in SRM 3233 Fortified Breakfast Cereal (method comparison data summary view – digestion method). In this view, individual laboratory data are plotted with the individual laboratory standard deviation (error bars). The color of the data point represents the sample preparation (digestion) procedure employed. Data points that are unfilled represent laboratories that only reported a single value for that analyte and therefore were not included in the consensus mean. The black solid line represents the consensus mean, and the black dotted lines represent the consensus variability calculated as one standard deviation about the consensus mean. The gray shaded region represents the target zone for "acceptable" performance, which encompasses the NIST certified value bounded by twice its uncertainty (U_{95}).

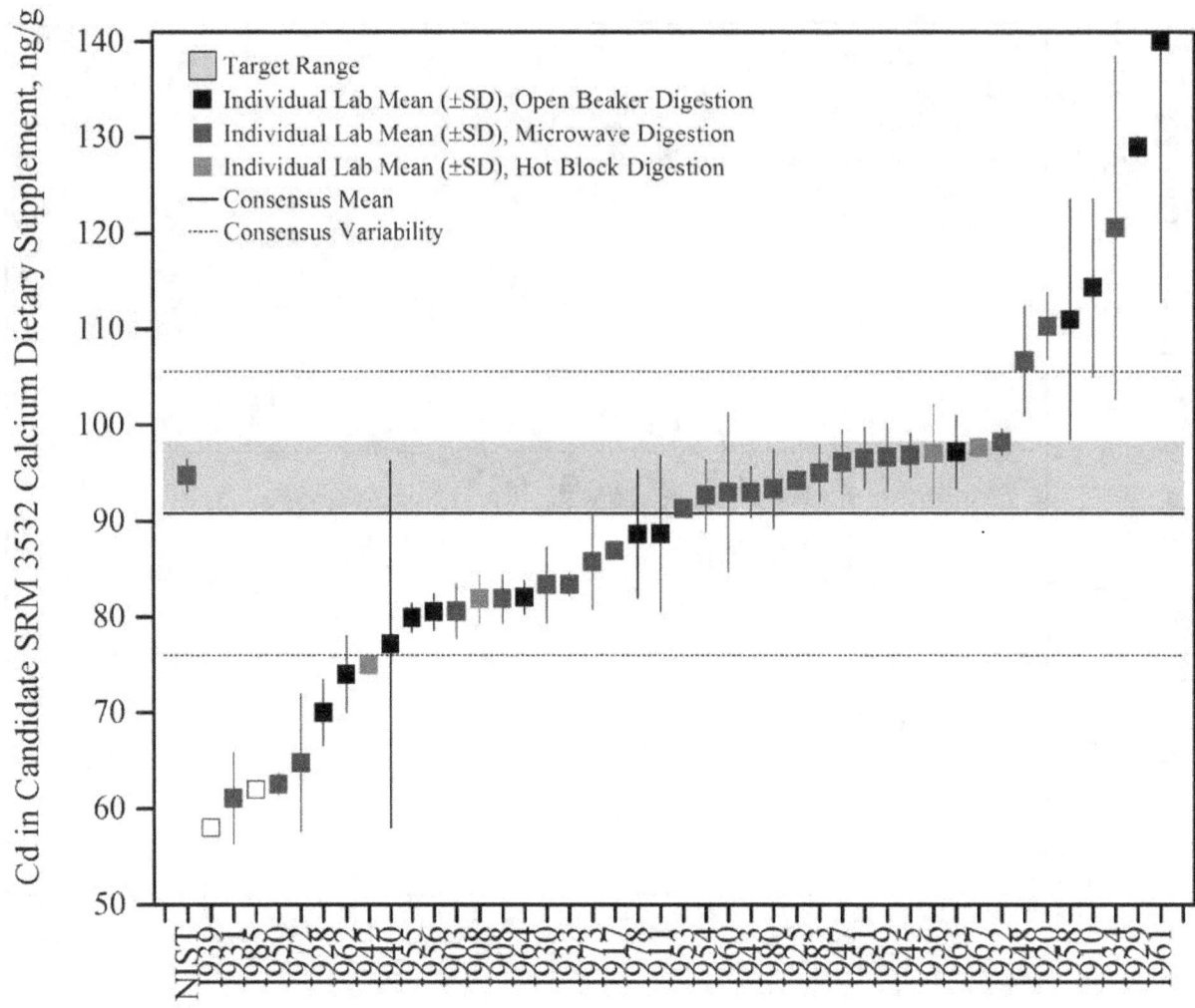

Figure 11. Cadmium in candidate SRM 3532 Calcium Dietary Supplement (method comparison data summary view – digestion method). In this view, individual laboratory data are plotted with the individual laboratory standard deviation (error bars). The color of the data point represents the sample preparation (digestion) procedure employed. Data points that are unfilled represent laboratories that only reported a single value for that analyte and therefore were not included in the consensus mean. The black solid line represents the consensus mean, and the black dotted lines represent the consensus variability calculated as one standard deviation about the consensus mean. The gray shaded region represents the target zone for "acceptable" performance, which encompasses the NIST value determined by ID-ICP-MS, bounded by twice the estimated uncertainty observed for six duplicate measurements.

28

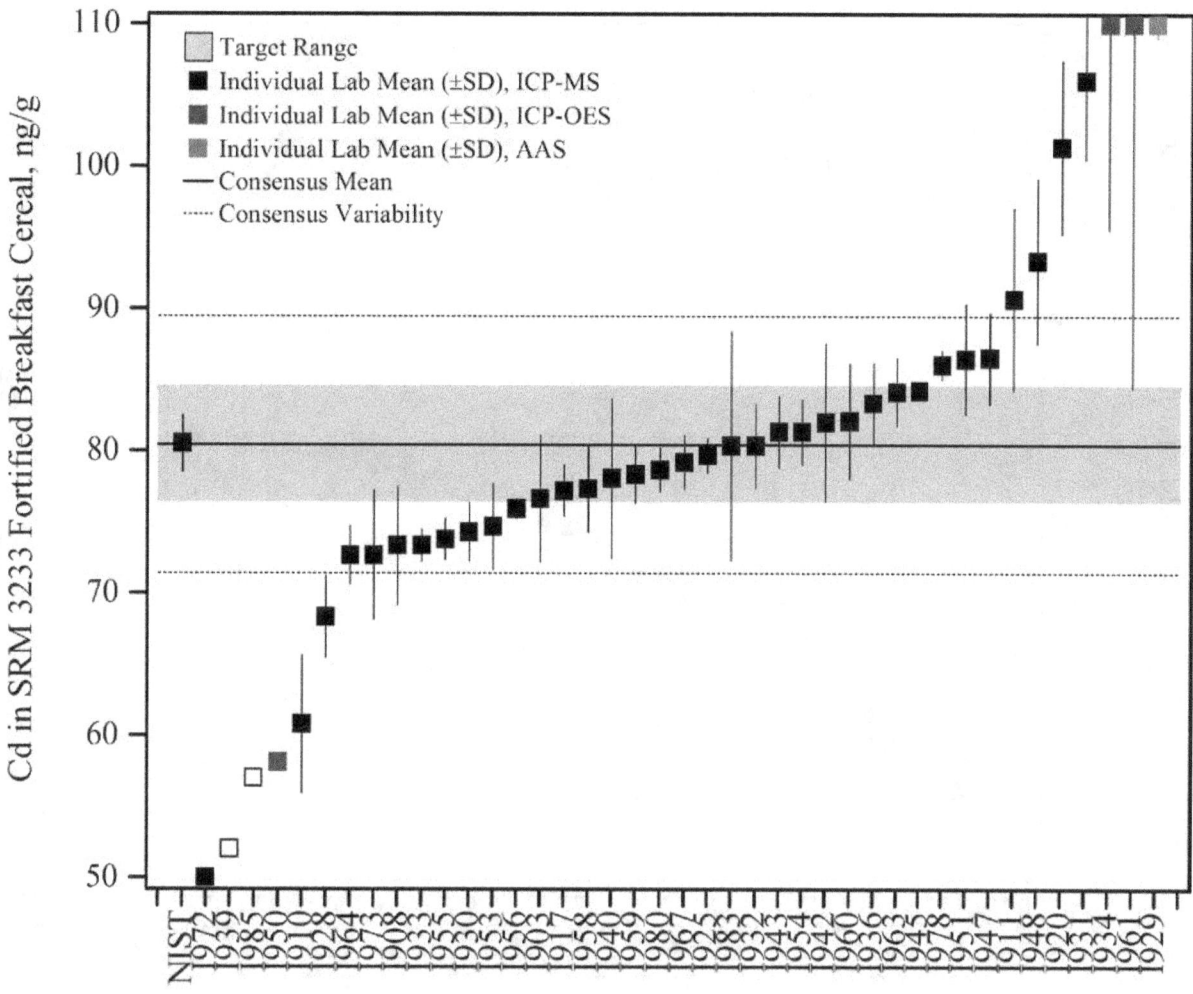

Figure 12. Cadmium in 3233 Fortified Breakfast Cereal (method comparison data summary view – instrumental method). In this view, individual laboratory data are plotted with the individual laboratory standard deviation (error bars). The color of the data point represents the instrumental method employed. Data points that are unfilled represent laboratories that only reported a single value for that analyte and therefore were not included in the consensus mean. The black solid line represents the consensus mean, and the black dotted lines represent the consensus variability calculated as one standard deviation about the consensus mean. The gray shaded region represents the target zone for "acceptable" performance, which encompasses the NIST certified value bounded by twice its uncertainty (U_{95}).

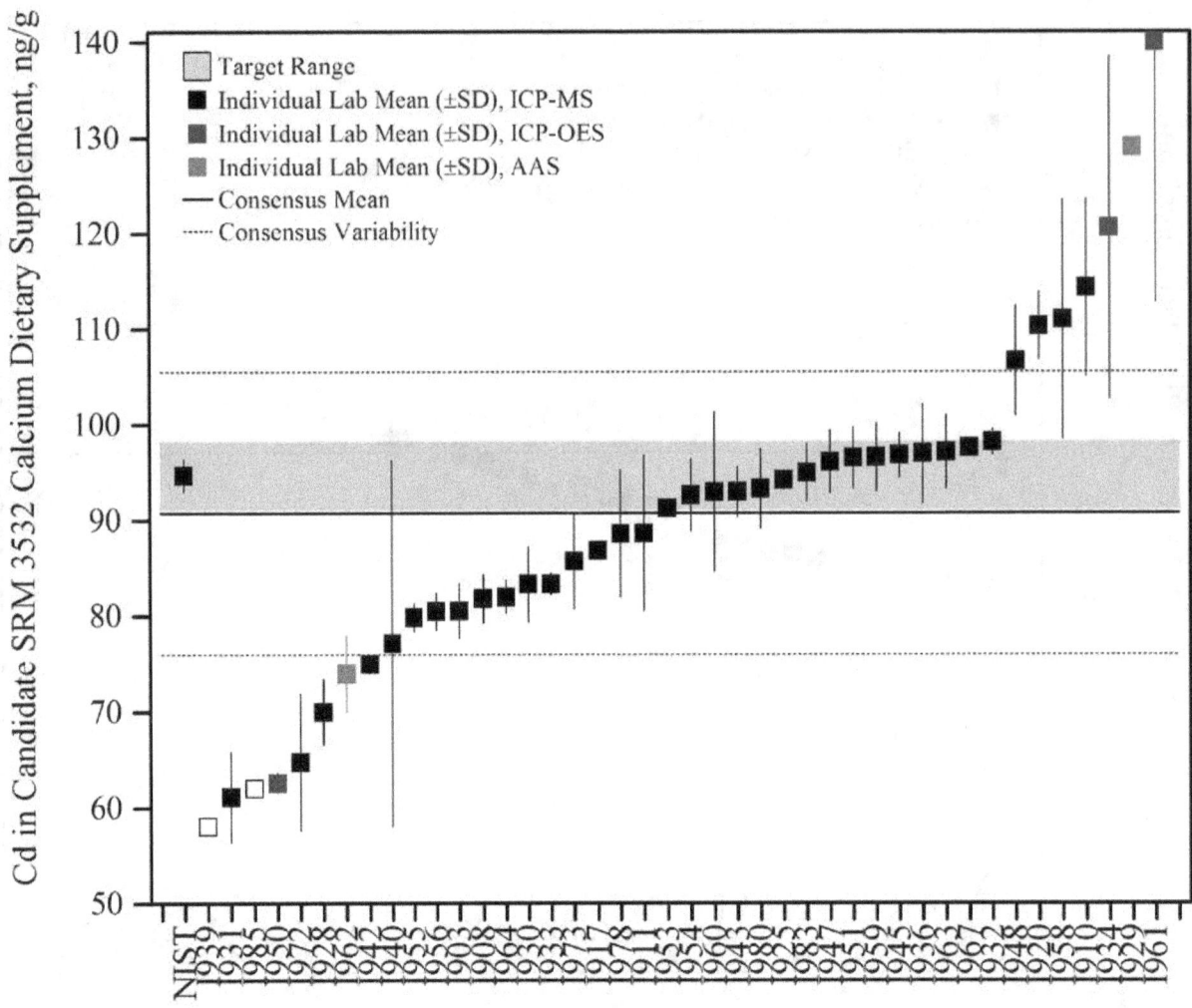

Figure 13. Cadmium in candidate SRM 3532 Calcium Dietary Supplement (method comparison data summary view – instrumental method). In this view, individual laboratory data are plotted with the individual laboratory standard deviation (error bars). The color of the data point represents the instrumental method employed. Data points that are unfilled represent laboratories that only reported a single value for that analyte and therefore were not included in the consensus mean. The black solid line represents the consensus mean, and the black dotted lines represent the consensus variability calculated as one standard deviation about the consensus mean. The gray shaded region represents the target zone for "acceptable" performance, which encompasses the NIST value determined by ID-ICP-MS, bounded by twice the estimated uncertainty observed for six duplicate measurements.

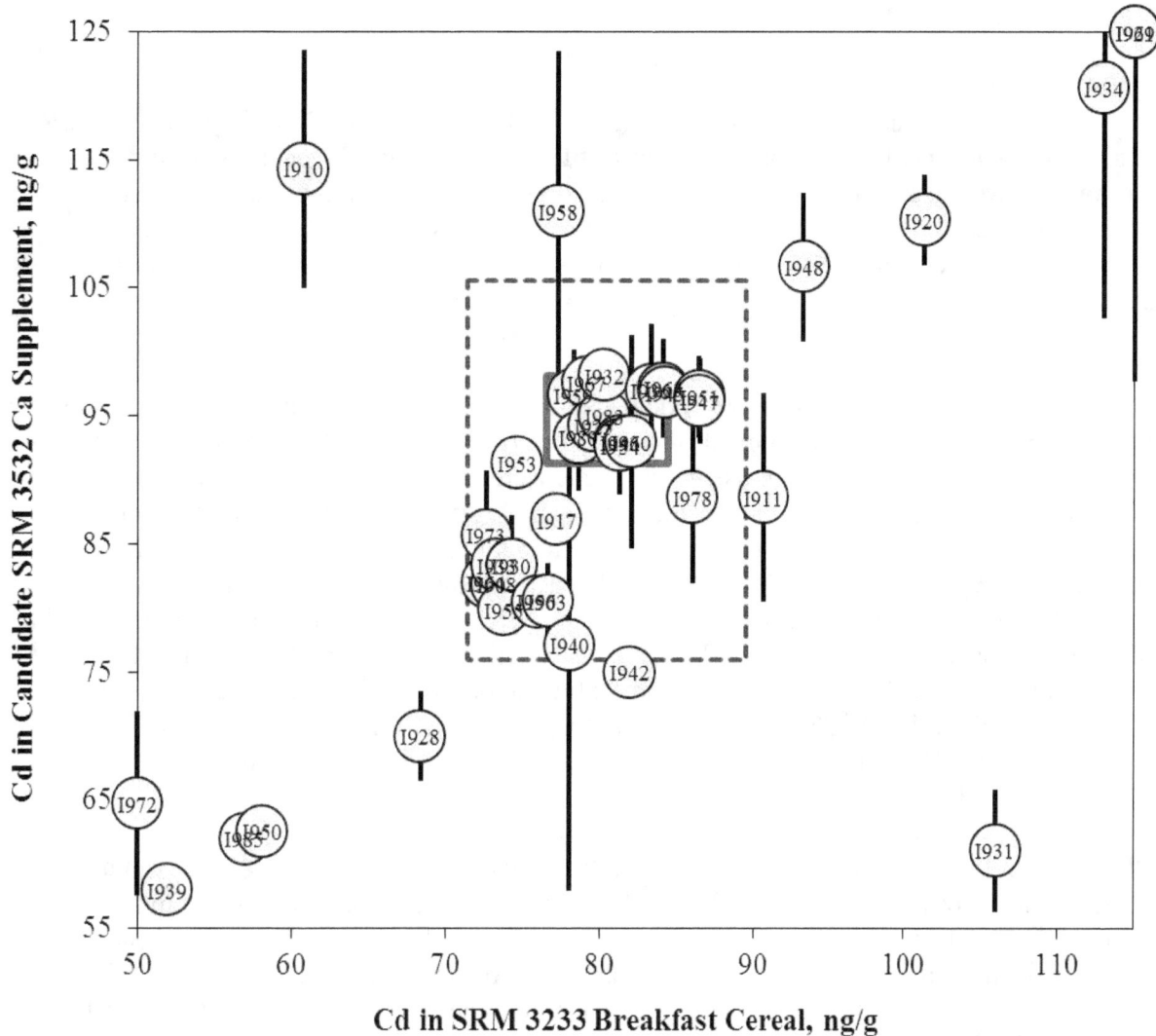

Figure 14. Cadmium in candidate SRM 3532 Calcium Dietary Supplement and SRM 3233 Fortified Breakfast Cereal (sample/control comparison view). In this view, the individual laboratory results for the control (SRM 3233 Fortified Breakfast Cereal) with a certified value for the analyte are compared to the results for a sample (candidate SRM 3532 Calcium Dietary Supplement). The error bars represent the individual laboratory standard deviation. The solid red lines represent the target zone for the control (x-axis) and the unknown sample (y-axis). The dotted blue box represents the consensus zone for the control (x-axis) and the unknown sample (y-axis).

VITAMIN B$_5$ IN FOODS

Study Overview
In this study, participants were provided with two NIST SRMs, SRM 3234 Soy Flour and SRM 3287 Blueberries, neither of which has been fortified with vitamin B$_5$ (pantothenic acid). Participants were asked to use in-house analytical methods to determine the mass fractions of vitamin B$_5$ in each of the matrices and report values on an as-received basis. Participants were asked to report the vitamin B$_5$ content as pantothenic acid; NIST values are reported as pantothenic acid.

Sample Information
Soy Flour. Participants were provided one packet containing approximately 15 g of defatted soy flour. The flour was heat-sealed inside 0.1 mm (4 mil) polyethylene bags, which were then sealed inside Mylar bags. Before use, participants were instructed to thoroughly mix and homogenize the contents of the packet and use a sample size of at least 2 g. Participants were asked to store the soy flour at controlled room temperature, 10 °C to 30 °C, prepare three samples, and report three values from the single packet provided. Approximate analyte levels were not provided to participants in the study. The NIST certified value for pantothenic acid in SRM 3234 was determined using acidic solvent extraction followed by ID-LC-MS/MS with confirmation using data from external collaborating laboratories. The certified value for pantothenic acid in SRM 3234 is (11.45 ± 0.12) mg/kg on a dry-mass basis. After adjustment for moisture content of the material of 6.13 %, the as-received target value for pantothenic acid in SRM 3234 is (10.75 ± 0.11) mg/kg.

Blueberries. Participants were provided with three packets, each containing approximately 5 g of freeze-dried, powdered blueberries. The blueberries were blended, aliquotted, and heat-sealed inside nitrogen-flushed 0.1 mm (4 mil) polyethylene bags, which were then sealed inside nitrogen-flushed aluminized plastic bags along with two packets of silica gel each. Before use, participants were instructed to thoroughly mix and homogenize the contents of the packet and use a sample size of at least 2.5 g. Participants were also notified that this material was packaged as a powder, but that over time the powder may become a solid mass. For hardened samples, participants were instructed to remove an appropriate test portion using a knife. Participants were asked to report a single value from each packet provided and to store the blueberries at controlled room temperature, 10 °C to 30 °C. Approximate analyte levels were not provided to participants prior to the study. The NIST certified value for pantothenic acid in SRM 3287 was determined using acidic solvent extraction followed by ID-LC-MS in combination with data from external collaborating laboratories. The certified value of pantothenic acid in SRM 3287 is (3.36 ± 0.19) mg/kg. After adjustment for moisture content of the material of 1.41 %, the as-received target value for pantothenic acid in SRM 3287 is (3.31 ± 0.19) mg/kg.

Study Results
- Thirty-three laboratories enrolled in this exercise and received samples, and thirteen laboratories reported results for the soy flour and/or blueberries (39 % participation).
- For both materials, the consensus ranges were very wide. For the soy flour, the consensus mean was higher than the NIST target range, while the consensus mean for the blueberries was contained within the NIST target range (**Figure 15** and **Figure 16**).

- The dispersion of the data could be a result of challenges in completely extracting the pantothenic acid from the samples or from chromatographic coelutions.
- In the soy flour, nine of the thirteen laboratories (69 %) reported values that were reasonably close to the target range. Three of the remaining four laboratories reported values that were significantly higher than the target range (10 times higher and almost 300 times higher). This could indicate an interference in the analytical method (LC-absorbance with external standard calibration) caused by matrix components. When using a low wavelength (205 nm to 210 nm) to detect pantothenic acid (a molecule without a strong chromophore), the method will be highly susceptible to matrix interferences. More information is needed about the analytical methods to draw more conclusions.
- In the blueberries, eight of the eleven laboratories (73 %) reported values that were reasonably close to the target range. Two of the remaining three laboratories reported values that were significantly higher than the target range (150 to 200 times higher). These were the same laboratories that reported high values for the soy flour, indicating a potential interference in the analytical method or possibly a calibration error.
- One laboratory reported values that were 10 times lower than the target value for the soy flour and 35 times lower for the blueberries. This could be the result of ion suppression in the MS, as this laboratory reported using LC-MS with an external standard calibration approach. For accurate quantitation from matrix-based samples, the use of isotope dilution for internal standard calibration is critical.
- In general, the analytical approach used did not correlate with any trend in the data. In this case, variability in the data is more likely related to the combination of sample preparation, instrumental method, and calibration approach, as any method must be careful to account for interferences. A larger data set and more information from participants is necessary to draw any strong correlations between method and result.

Technical Recommendations

The following are recommendations based on results provided by the participants in this study.
- No analytical method was identified as being exceptionally good or problematic. When using LC-absorbance for a molecule like pantothenic acid without a chromophore, care must be taken to remove matrix interferences. The same is true when using LC-MS, as matrix components can cause ion suppression leading to results that are biased low unless an isotopically labeled analog is used for internal standard calibration.

Table 7. Individual data table (NIST) for vitamin B_5 (pantothenic acid) in foods.

National Institute of Standards & Technology

Exercise I – October 2012 – Pantothenic Acid

	Lab Code:	NIST	1. Your Results				2. Community Results			3. Target	
Analyte	Sample	Units	x_i	s_i	Z_{comm}	Z_{NIST}	N	x^*	s^*	x_{NIST}	U_{95}
B_5	Soy Flour	µg/g	10.7	0.10	2.3	-0.4	13	18.6	15.7	10.7	0.1
B_5	Blueberries	µg/g	3.31	0.19	-0.1	0.0	11	3.49	3.08	3.31	0.19

x_i Mean of reported values

s_i Standard deviation of reported values

Z_{comm} Z-score with respect to community consensus

Z_{NIST} Z-score with respect to NIST value

N   Number of quantitative values reported

x^* Robust mean of reported values

s^* Robust standard deviation

x_{NIST} NIST-assessed value

U_{95} ±95% confidence interval about the assessed value or standard deviation (s_{NIST})

34

Table 8. Data summary table for vitamin B$_5$ (pantothenic acid) in foods.

		Pantothenic Acid									
		SRM 3234 Soy Flour (μg/g)					SRM 3287 Blueberries (μg/g)				
	Lab	A	B	C	Avg	SD	A	B	C	Avg	SD
Individual Results	NIST				10.7	0.1				3.31	0.19
	I903	11.1	11.0	10.8	11.0	0.2	4.20	3.90	4.20	4.10	0.17
	I905										
	I907										
	I910										
	I911										
	I914										
	I916										
	I919	11.6	12.0	11.8	11.8	0.2	4.61	3.35	4.42	4.13	0.68
	I924										
	I928										
	I931	13.3	12.5	12.3	12.7	0.5	1.71	1.41	1.48	1.53	0.16
	I932	16.1	17.5	19.8	17.8	1.9	1.10	5.06	0.83	2.33	2.37
	I933										
	I935										
	I936	13.0	13.0	13.0	13.0	0.0	1.20	1.50	1.50	1.40	0.17
	I937	210.0	226.0	214.0	216.7	8.3					
	I938										
	I940	120.9	133.2	142.3	132.1	10.8	540.17	510.45	462.82	504.48	39.02
	I941	12.4	14.6	12.4	13.1	1.2	4.82	5.53	4.87	5.07	0.40
	I946										
	I950										
	I958										
	I959	12.2	11.6	11.0	11.6	0.6	2.53	2.37	2.37	2.42	0.09
	I961										
	I963	2904.6	2846.0	2907.4	2886.0	34.7	642.17	656.83	652.90	650.63	7.58
	I971										
	I974										
	I975										
	I976										
	I978	1.1	0.9	1.0	1.0	0.1	0.10	0.10	0.09	0.09	0.01
	I979										
	I980	11.8	11.4	11.6	11.6	0.2	1.30	1.00	1.00	1.10	0.17
	I983										
Community Results		Consensus Mean			14.7		Consensus Mean			3.41	
		Consensus Standard Deviation			6.0		Consensus Standard Deviation			2.92	
		Maximum			2886.0		Maximum			650.63	
		Minimum			1.0		Minimum			0.09	
		N			12		N			11	

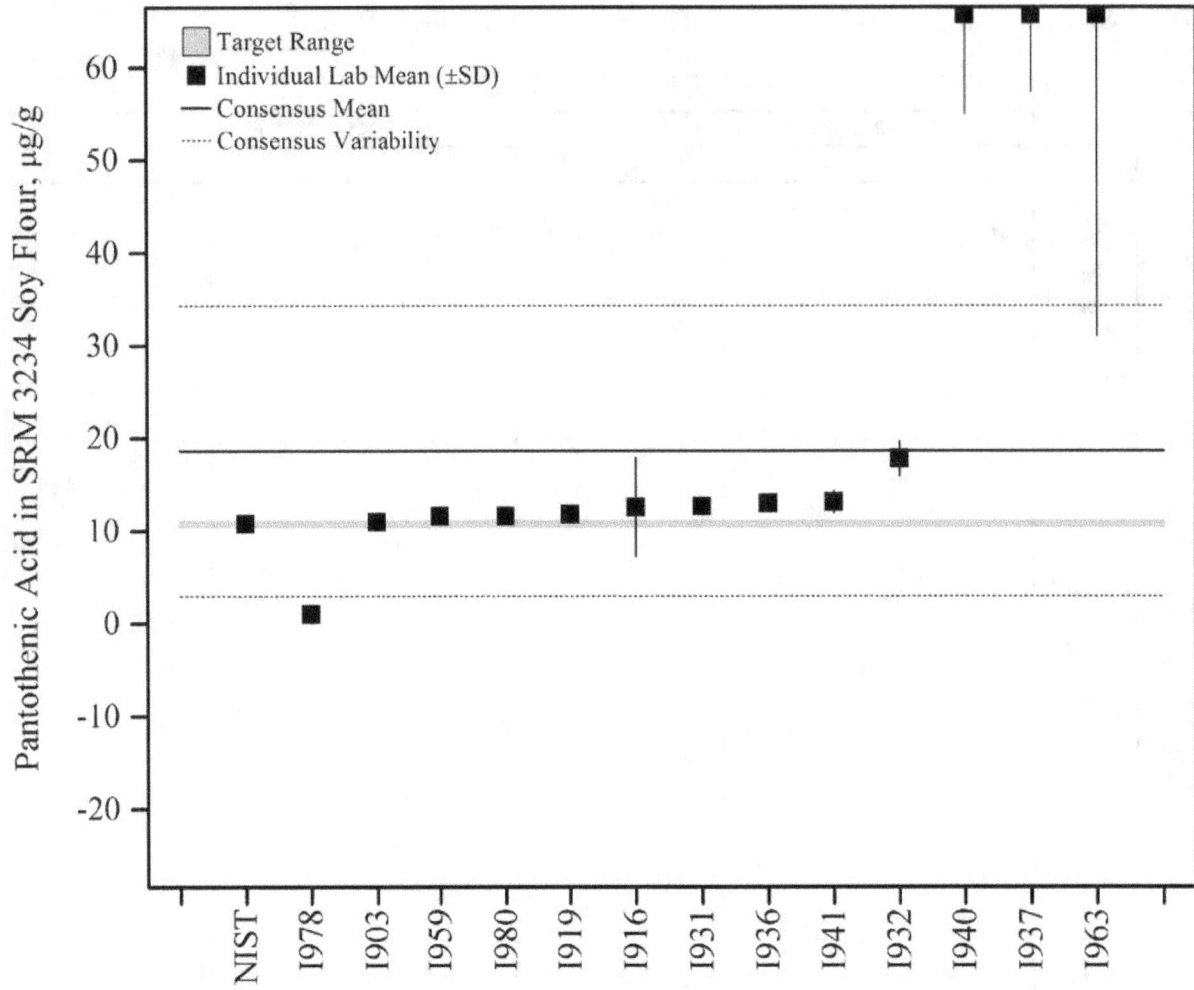

Figure 15. Vitamin B₅ (pantothenic acid) in SRM 3234 Soy Flour (data summary view). In this view, individual laboratory data are plotted with the individual laboratory standard deviation (error bars). The black solid line represents the consensus mean, and the black dotted lines represent the consensus variability calculated as one standard deviation about the consensus mean. The gray shaded region represents the target zone for "acceptable" performance, which encompasses the NIST certified value determined by ID-LC-MS/MS and external collaborating laboratories bounded by twice the uncertainty (U_{95}).

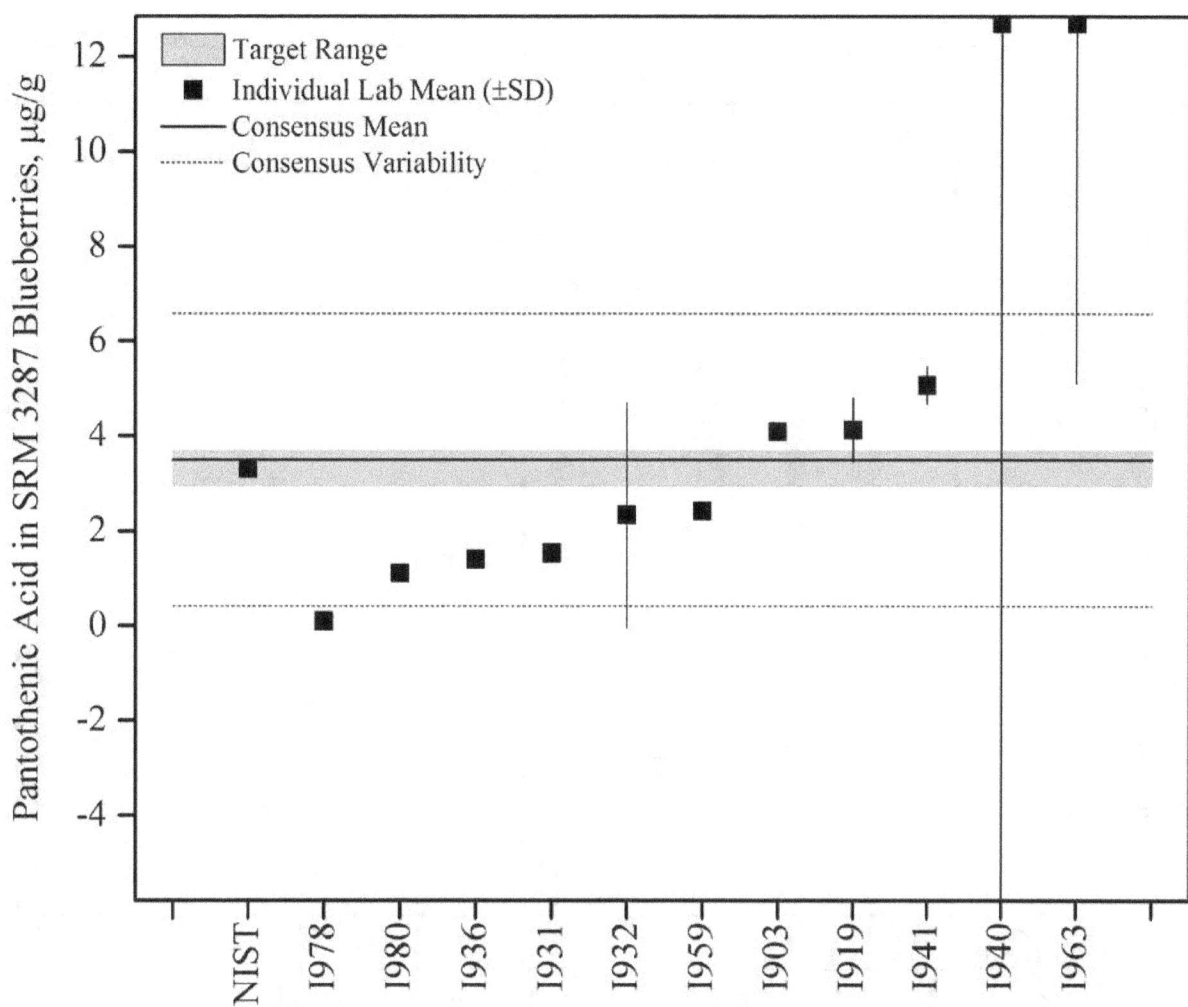

Figure 16. Vitamin B₅ (pantothenic acid) in SRM 3287 Blueberries (data summary view). In this view, individual laboratory data are plotted with the individual laboratory standard deviation (error bars). The black solid line represents the consensus mean, and the black dotted lines represent the consensus variability calculated as one standard deviation about the consensus mean. The gray shaded region represents the target zone for "acceptable" performance, which encompasses the NIST certified value determined by ID-LC-MS and external collaborating laboratories bounded by twice the uncertainty (U_{95}).

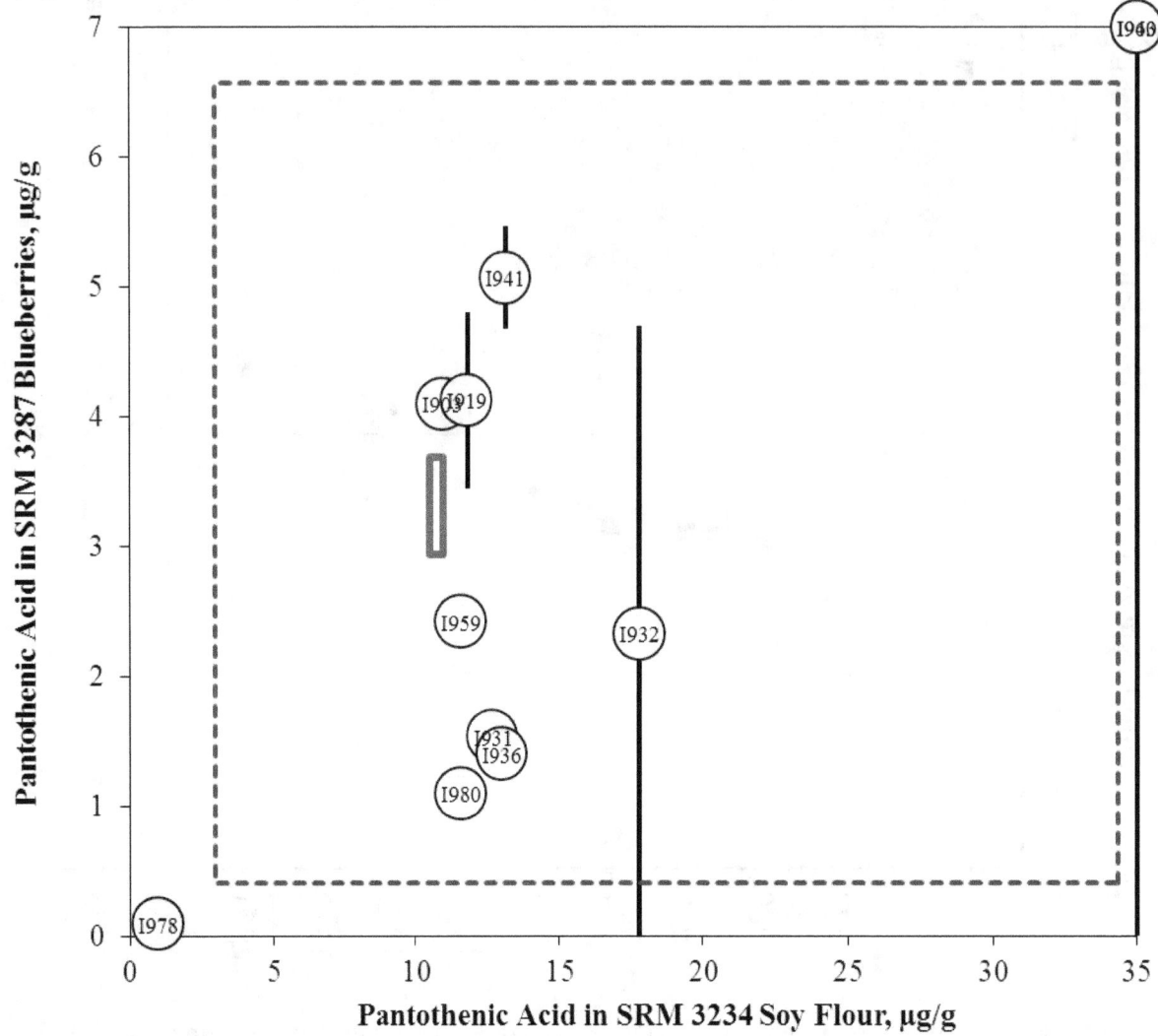

Figure 17. Vitamin B₅ (pantothenic acid) in SRM 3234 Soy Flour and SRM 3287 Blueberries (sample/sample comparison view). In this view, the individual laboratory results for one sample (SRM 3287 Blueberries) with a certified value for the analyte are compared to the results for a second sample (SRM 3234 Soy Flour). The error bars represent the individual laboratory standard deviation. The solid red lines represent the target zone for the control (x-axis) and the unknown sample (y-axis). The dotted blue box represents the consensus zone for the control (x-axis) and the unknown sample (y-axis).

VITAMIN A IN FOODS AND SUPPLEMENTS

<u>Study Overview</u>
In this study, participants were provided with SRM 3280 Multivitamin/Multielement Tablets and a whole egg powder. Participants were asked to use in-house analytical methods to determine the mass fractions of vitamin A (as retinol, retinyl acetate, and retinyl palmitate) in each of the matrices and report values on an as-received basis.

<u>Sample Information</u>
Multivitamin/Multielement Tablets. Participants were provided with one bottle containing 30 multivitamin/multielement tablets. Before use, participants were instructed to grind all tablets together, mix the resulting powder thoroughly, and use a sample size of at least 0.6 g. Participants were asked to store the material at controlled room temperature, 10 °C to 30 °C, prepare three samples, and report three values from the single bottle provided. Approximate analyte levels were not provided to participants prior to the study. The NIST reference values and uncertainties for vitamin A in SRM 3280 were determined by LC-MS following solvent extraction and are reported in the table below, both on a dry-mass basis and after correction for moisture of the material (1.37 %).

Egg powder. Participants were provided with one packet containing approximately 10 g of whole egg powder. The material is a free-flowing, fine powder prepared from USDA-inspected whole eggs. Before use, participants were instructed to mix thoroughly the contents of the packet and use a sample size of at least 1 g. Participants were asked to store the material at controlled room temperature, 10 °C to 30 °C, prepare three samples, and report three values from the single packet provided. Approximate analyte levels were not provided to participants prior to the study, and NIST-assessed values and uncertainties were not determined for the whole egg powder.

<u>Study Results</u>
- Thirty-seven laboratories enrolled in this exercise and received samples, and 19 laboratories reported results for at least one of the samples (51 % participation).
- NIST target values are available for retinol equivalents and retinyl acetate in the multivitamin sample.
  - The consensus mean for retinol and retinyl acetate were within the target range.
  - The consensus ranges were acceptable for both compounds in the multivitamin sample (19 % for both compounds).
  - Two laboratories reported values for retinyl palmitate in the multivitamin sample. The value from one laboratory appeared to be a conversion of the measured mass fraction of retinyl acetate to retinyl palmitate using the relative molecular masses of the compounds.
- NIST target values are not available for retinol in the egg powder sample. The consensus range for retinol in the egg powder was quite wide (83 % RSD).
- Ten laboratories (53 %) reported using saponification followed by extraction, while nine laboratories (47 %) reported using solvent extraction to prepare samples.
- A majority of laboratories (95 %) used LC-absorbance for analysis. One laboratory reported using spectrophotometry.

- All laboratories reported using an external standard approach to calibration.

Technical Recommendations

The following recommendations are based on results provided by the participants in this study.

- It is important to determine that saponification methods are appropriate for a given sample. Conditions that are too extreme may result in degradation of the analyte of interest and conditions that are too gentle may not fully extract the analyte of interest. In future exercises, more survey information from participants will be collected about saponification to help aid in making recommendations.
- Always be certain that calibrants match the measured analyte (e.g., do not measure retinyl acetate with a retinol calibrant).
- Due to the nature of the calibrant materials, a spectrophotometric determination of calibrant concentration is essential for accurate measurements.

Table 9. Individual data table (NIST) for vitamin A in foods and supplements.

National Institute of Standards & Technology

Exercise I – October 2012 – Vitamin A

Analyte	Sample	Units	1. Your Results				2. Community Results			3. Target	
Lab Code:		NIST	x_i	s_i	Z_{comm}	Z_{NIST}	N	x*	s*	x_{NIST}	U_{95}
Retinol	Vitamin	µg/g	438	45	-0.1	0.0	9	447	85	438	45
Retinol	Egg Powder	µg/g					9	1.600	1.330		
Retinyl Acetate	Vitamin	µg/g	502	52	0.5	0.0	13	460	88	502	52
Retinyl Acetate	Egg Powder	µg/g					2	3.570	1.300		
Retinyl Palmitate	Vitamin	µg/g					2	392	464		
Retinyl Palmitate	Egg Powder	µg/g					1				

x_i Mean of reported values

s_i Standard deviation of reported values

Z_{comm} Z-score with respect to community consensus

Z_{NIST} Z-score with respect to NIST value

N Number of quantitative values reported

x* Robust mean of reported values

s* Robust standard deviation

x_{NIST} NIST-assessed value

U_{95} ±95% confidence interval about the assessed value or standard deviation (s_{NIST})

41

Table 10. Data summary table for retinol in foods.

	Lab	SRM 3280 Multivitamin Tablet (µg/g)					Whole Egg Powder (µg/g)				
		A	**B**	**C**	**Avg**	**SD**	**A**	**B**	**C**	**Avg**	**SD**
Individual Results	NIST				438	45					
	I901	387	410	409	402	13					
	I903	357	355	338	350	10	0.760	0.721	0.755	0.745	0.021
	I905						0.600	0.500	0.400	0.500	0.100
	I907										
	I910										
	I911										
	I914										
	I915										
	I916										
	I919	408	444	406	419	21	0.622	0.655	0.675	0.651	0.027
	I922										
	I924	425	408	385	406	20	2.100	1.770	1.250	1.707	0.429
	I928										
	I929	560	543	524	542	18	25.000	28.000	23.000	25.333	2.517
	I932										
	I933										
	I936										
	I937	434	451	433	439	10					
	I938										
	I940										
	I946										
	I949										
	I950										
	I958										
	I959	774	774	726	758	28	3.200	3.000	3.200	3.133	0.115
	I961										
	I963										
	I965										
	I971										
	I974										
	I975										
	I977										
	I978	400	304	409	371	58	2.874	2.001	2.351	2.409	0.439
	I979										
	I980						0.763	0.712	0.810	0.762	0.049
	I983										
	I984										
Community Results		Consensus Mean			449		Consensus Mean			1.73	
		Consensus Standard Deviation			94		Consensus Standard Deviation			1.47	
		Maximum			758		Maximum			25.33	
		Minimum			350		Minimum			0.50	
		N			8		N			8	

Table 11. Data summary table for retinyl acetate in foods.

	Lab	SRM 3280 Multivitamin Tablet (µg/g)					Whole Egg Powder (µg/g)				
		A	B	C	Avg	SD	A	B	C	Avg	SD
Individual Results	NIST				502	52					
	I901										
	I903	472	468	445	462	15					
	I905	452	486	465	468	17					
	I907										
	I910	490	508	512	503	12					
	I911										
	I914										
	I915										
	I916										
	I919	468	509	466	481	24					
	I922										
	I924										
	I928	399	413	404	405	7					
	I929										
	I932	501	514	505	507	7					
	I933	1670	1700	1660	1677	21					
	I936										
	I937										
	I938										
	I940	342	343	343	342	0	5.147	4.202	3.787	4.379	0.697
	I946										
	I949	475	456	453	461	12					
	I950										
	I958	700	789	760	750	45					
	I959										
	I961										
	I963	807	751	793	784	29					
	I965										
	I971										
	I974										
	I975										
	I977										
	I978	459	349	469	426	67	3.296	2.294	2.696	2.762	0.504
	I979										
	I980	384	384	357	375	16					
	I983										
	I984										

Community Results	Consensus Mean	493	Consensus Mean	3.57
	Consensus Standard Deviation	120	Consensus Standard Deviation	1.30
	Maximum	1677	Maximum	4.38
	Minimum	342	Minimum	2.76
	N	13	N	2

Table 12. Data summary table for retinyl palmitate in foods.

	Lab	SRM 3280 Multivitamin Tablet (µg/g)					Whole Egg Powder (µg/g)				
		A	B	C	Avg	SD	A	B	C	Avg	SD
Individual Results	NIST										
	I901										
	I903										
	I905										
	I907										
	I910										
	I911										
	I914										
	I915										
	I916										
	I919										
	I922										
	I924										
	I928										
	I929										
	I932										
	I933										
	I936										
	I937										
	I938										
	I940										
	I946										
	I949										
	I950										
	I958										
	I959										
	I961										
	I963	85	144	78	102	36					
	I965										
	I971										
	I974										
	I975										
	I977										
	I978	734	558	750	681	107	5.27	3.67	4.31	4.42	0.81
	I979										
	I980										
	I983										
	I984										
Community Results		Consensus Mean			392		Consensus Mean				
		Consensus Standard Deviation			464		Consensus Standard Deviation				
		Maximum			681		Maximum			4.42	
		Minimum			102		Minimum			4.42	
		N			2		N			1	

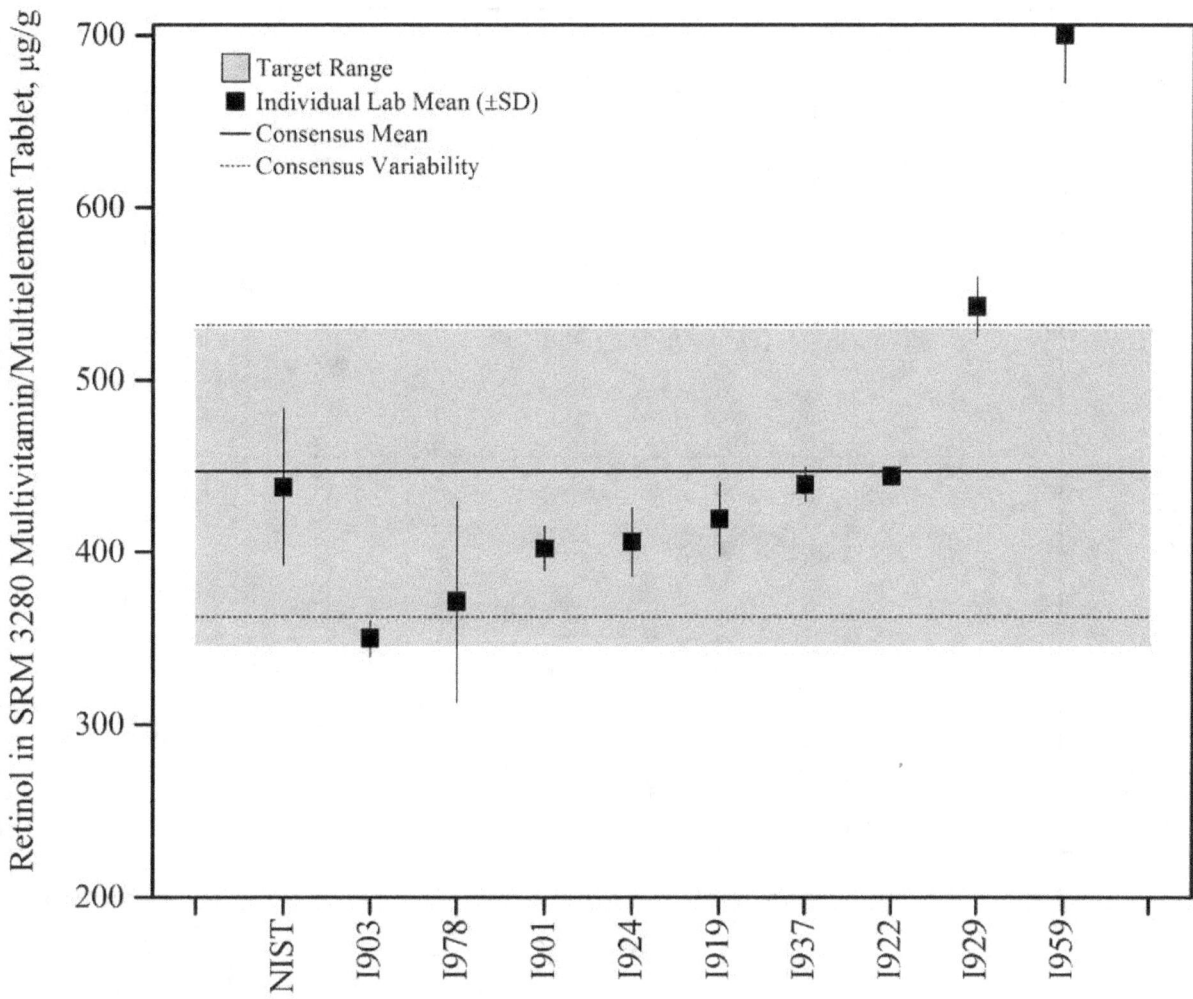

Figure 18. Retinol in SRM 3280 Multivitamin/Multielement Tablet (data summary view). In this view, individual laboratory data are plotted with the individual laboratory standard deviation (error bars). Data points that are unfilled represent laboratories that only reported a single value for that analyte and therefore were not included in the consensus mean. The black solid line represents the consensus mean, and the black dotted lines represent the consensus variability calculated as one standard deviation about the consensus mean. The gray shaded region represents the target zone for "acceptable" performance, which encompasses the NIST reference value determined by LC-MS (measured as retinyl acetate, expressed as retinol equivalents) bounded by twice the uncertainty (U_{95}).

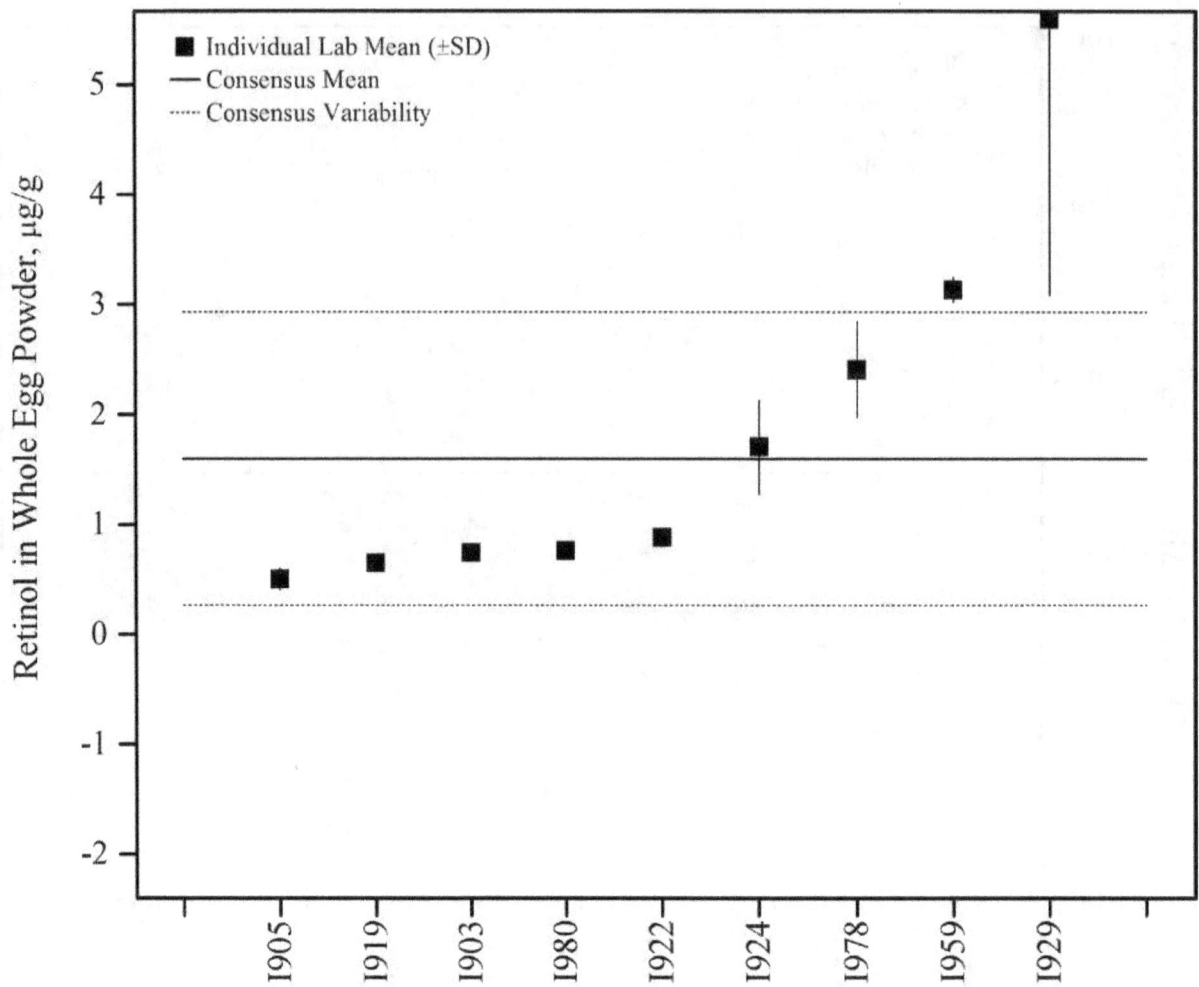

Figure 19. Retinol in whole egg powder (data summary view). In this view, individual laboratory data are plotted with the individual laboratory standard deviation (error bars). The black solid line represents the consensus mean, and the black dotted lines represent the consensus variability calculated as one standard deviation about the consensus mean.

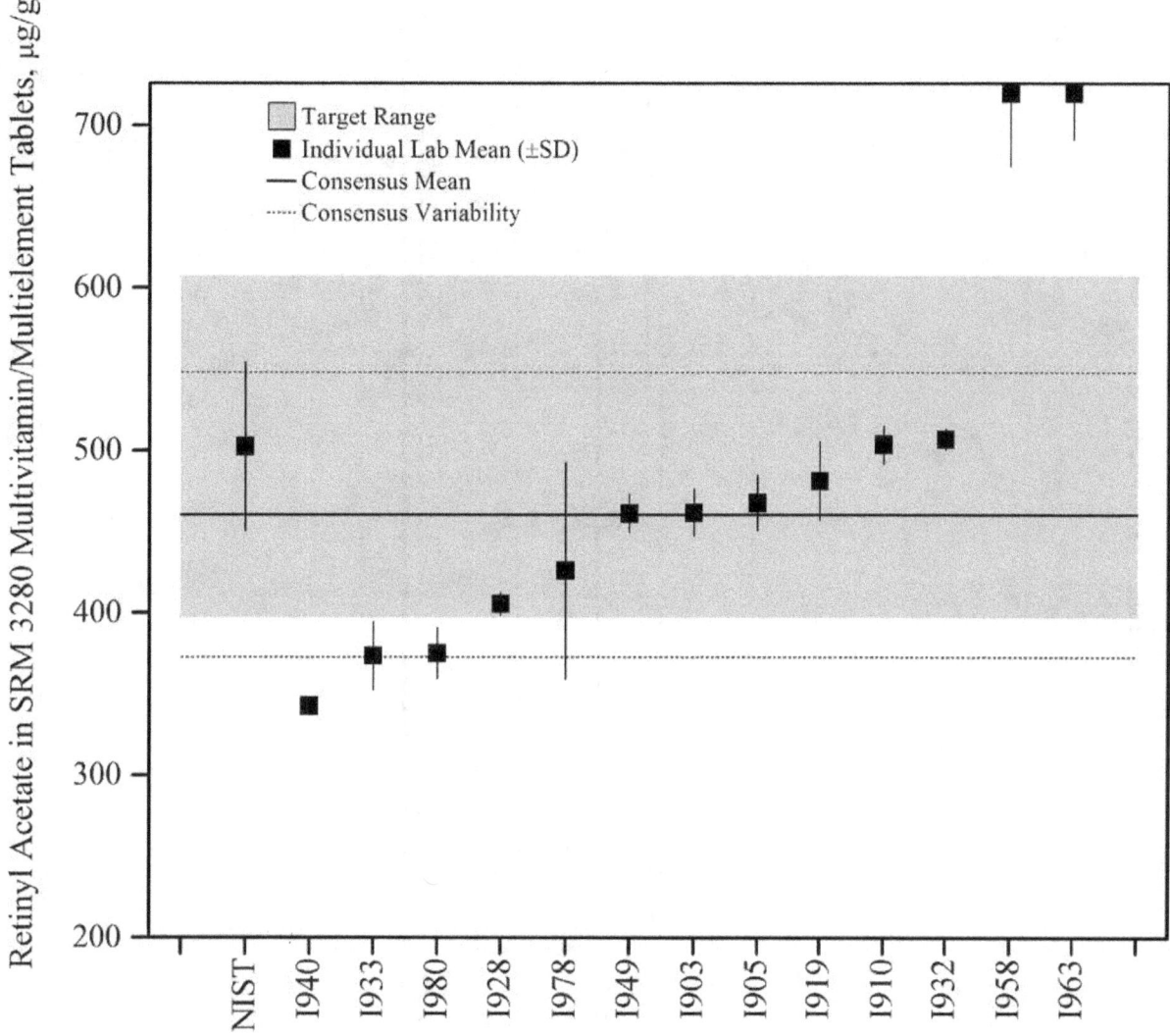

Figure 20. Retinyl acetate in SRM 3280 Multivitamin/Multielement Tablets (data summary view). In this view, individual laboratory data are plotted with the individual laboratory standard deviation (error bars). The black solid line represents the consensus mean, and the black dotted lines represent the consensus variability calculated as one standard deviation about the consensus mean. The gray shaded region represents the target zone for "acceptable" performance, which encompasses the NIST reference value determined by LC-MS bounded by twice the uncertainty (U_{95}).

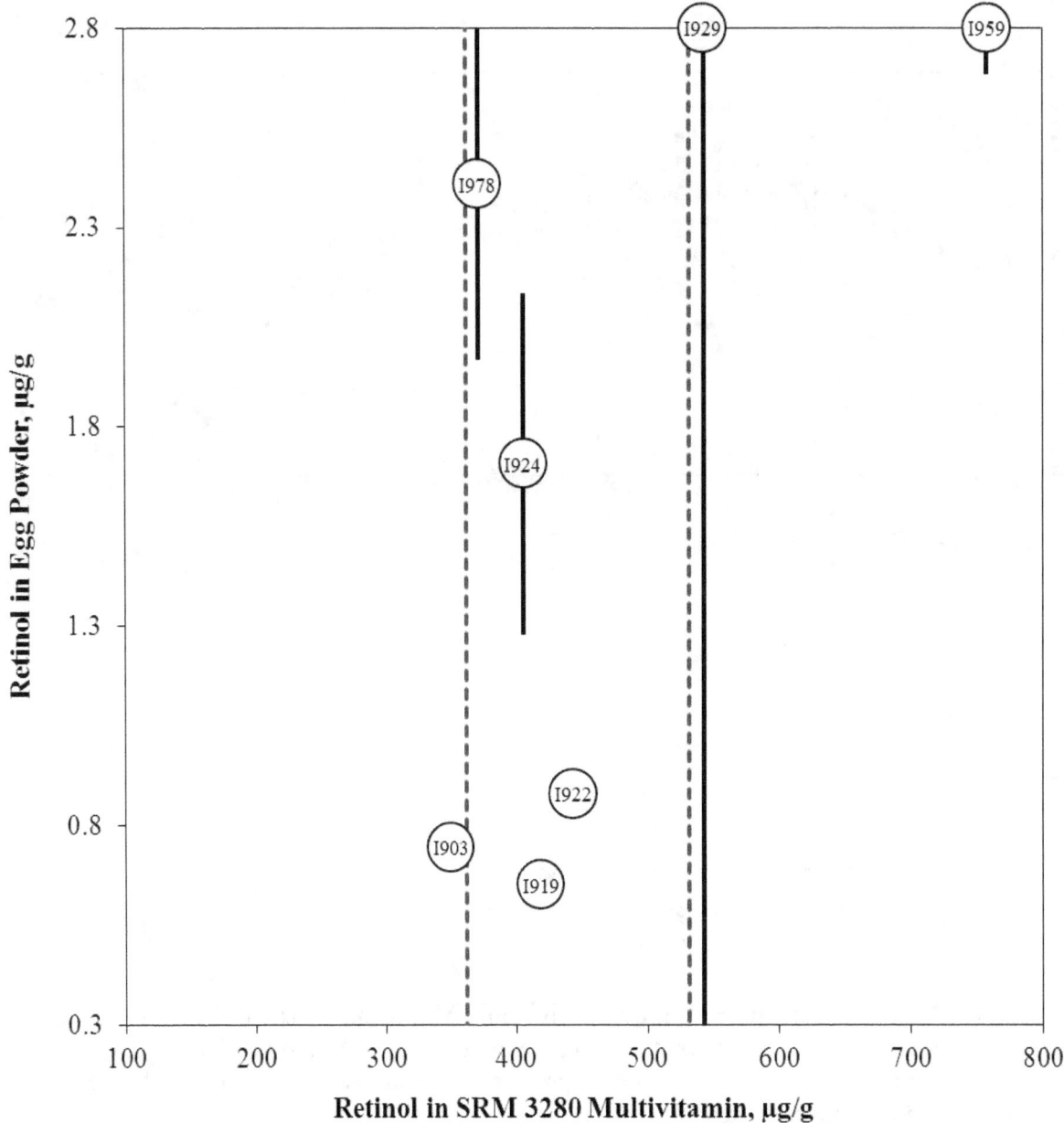

Figure 21. Retinol in whole egg powder and SRM 3280 Multivitamin/Multielement Tablets (sample/control comparison view). In this view, the individual laboratory results for the control (SRM 3280 Multivitamin/Multielement Tablets) with a reference value for the analyte are compared to the results for an unknown (whole egg powder). The error bars represent the individual laboratory standard deviation. The dotted blue box represents the consensus zone for the control (x-axis) and the unknown sample (y-axis).

CATECHINS IN GREEN TEA

Study Overview
In this study, participants were provided with two NIST SRMs, SRM 3255 *Camellia sinensis* (Green Tea) Extract and SRM 3254 *Camellia sinensis* (Green Tea) Leaves. Participants were asked to use in-house analytical methods to determine the mass fractions of seven catechins (catechin, epicatechin, epicatechin gallate, epigallocatechin, epigallocatechin gallate, gallocatechin, and gallocatechin gallate), as well as the total amount of catechins, in each of the matrices and report values on an as-received basis.

Sample Information
Green tea extract. Participants were provided with three packets, each containing approximately 1 g of an extract of green tea extract. The spray-dried extract of green tea leaves was heat-sealed inside nitrogen-flushed 0.1 mm (4 mil) polyethylene bags, which were then sealed inside aluminized plastic bags with two packets of silica gel. Before use, participants were instructed to thoroughly mix the contents of each packet and use a sample size of at least 50 mg. Participants were asked to store the extract at controlled room temperature, 10 °C to 30 °C, and report a single value from each packet. Approximate analyte levels were not provided to participants prior to the study. The NIST certified values in SRM 3255 were determined by LC-UV, LC-MS, and data from external collaborating laboratories. The certified values and their associated uncertainties, corrected for the moisture content of the material (3.13 %), are provided on an as-received basis in the table below.

Green tea leaves. Participants were provided with three packets, each containing approximately 3 g of green tea leaves. The ground green tea leaves were heat-sealed inside nitrogen-flushed 0.1 mm (4 mil) polyethylene bags, which were then sealed inside aluminized plastic bags with 2 packets of silica gel. Before use, participants were instructed to thoroughly mix the contents of the packet and use a sample size of at least 0.4 g. Participants were asked to store the material at controlled room temperature, 10 °C to 30 °C, and report a single value from each packet. Approximate analyte levels were not provided to participants prior to the study. The NIST certified values in SRM 3254 were determined by LC-UV, LC-MS, and data from external collaborating laboratories. The certified values and their associated uncertainties, corrected for the moisture content of the material (5.19 %), are provided on an as-received basis in the table below.

Analyte	Certified Mass Fraction in SRM 3255 (mg/g) (as-received basis)			Certified Mass Fraction in SRM 3254 (mg/g) (as-received basis)		
Catechin	8.88	±	0.90	0.958	±	0.389
Epicatechin	45.8	±	6.5	8.53	±	1.5
Epicatechin Gallate	97.2	±	7.6	12.0	±	1.1
Epigallocatechin	79.2	±	6.3	23.9	±	4.3
Epigallocatechin Gallate	409	±	18	49.3	±	2.1
Gallocatechin	21.3	±	1.6	2.28	±	1.0
Gallocatechin Gallate	37.8	±	1.9	0.939	±	0.20
Total Catechins	699	±	22	97.9	±	5.2

<u>Study Results</u>
- Forty-nine laboratories enrolled in this exercise and received samples, and twenty-eight laboratories reported results (57 % participation).
- The consensus means for catechin, epicatechin, epicatechin gallate, and epigallocatechin gallate in the extract were within the target range, with acceptable consensus ranges (9 % to 24 % RSD).
- The consensus mean for epigallocatechin was slightly below the target range, while the consensus means for gallocatechin and gallocatechin gallate were slightly above the target range. The consensus ranges were quite wide for all three (25 % to 72 % RSD).
- The consensus means for catechin, epicatechin, epicatechin gallate, epigallocatechin, and epigallocatechin gallate in the ground tea leaves were within the target range, with acceptable consensus ranges for epicatechin, epicatechin gallate, and epigallocatechin gallate (21 % to 23 % RSD). The consensus ranges for catechin and epigallocatechin were significantly wider (67 % RSD and 59 % RSD, respectively).
- The consensus means for gallocatechin and gallocatechin gallate were higher than the target range with wide consensus ranges (65 % to 112 % RSD).
- The consensus means for total catechins in both the extract and the leaves were within the target range, with acceptable consensus ranges (11 % and 24 % RSD, respectively).
- Laboratories that reported low values typically reported low values for all of the analytes in both matrices. The same is true for those laboratories reporting high values.
- Twenty-seven (96 %) of the laboratories reported using solvent extraction followed by LC-absorbance with external standard calibration. One laboratory reported using solvent extraction with LC-MS and external standard calibration.
- This study was previously conducted in Exercise E of the DSQAP (2010). The results for this study are significantly improved, with twice as many laboratories participating and more consistent results for nearly all of the individual catechins.

<u>Technical Recommendations</u>
The following recommendations are based on results provided by the participants in this study.
- Some laboratories (those reporting all high or low values) may have a calibration or sample preparation issue. Calibrant materials should be subjected to the same preparation procedure as the samples (derivatization, hydrolysis, etc.), and individual calibration standards should be used for each compound to improve accuracy.
- When sample preparation is extensive, an internal standard approach may be required to improve accuracy and precision.
- If an internal standard approach is used, it is best to add the internal standard at the earliest possible point (i.e. prior to extraction, saponification, and/or derivatization).

Table 13. Individual data table (NIST) for catechins in green tea.

National Institute of Standards & Technology

Exercise I – October 2012 – Catechins

Lab Code:	NIST		1. Your Results				2. Community Results			3. Target	
Analyte	Sample	Units	x_i	s_i	Z_{comm}	Z_{NIST}	N	x^*	s^*	x_{NIST}	U_{95}
Catechin	Extract	mg/g	8.88	0.90	-0.4	0.0	24	9.84	2.37	8.88	0.90
Catechin	Tea	mg/g	0.958	0.389	-0.6	0.0	20	1.54	1.04	0.958	0.389
Epicatechin	Extract	mg/g	45.8	6.5	0.5	0.0	26	43.2	5.8	45.8	6.5
Epicatechin	Tea	mg/g	8.53	1.52	0.6	0.0	25	7.57	1.6	8.53	1.5
Epicatechin Gallate	Extract	mg/g	97.2	7.6	0.2	0.0	25	95.1	13.4	97.2	7.6
Epicatechin Gallate	Tea	mg/g	12.0	1.1	0.0	0.0	24	11.9	2.7	12.0	1.1
Epigallocatechin	Extract	mg/g	79.2	6.3	0.5	0.0	24	62.9	34.8	79.2	6.3
Epigallocatechin	Tea	mg/g	23.9	4.3	0.3	0.0	23	20.6	12.2	23.9	4.3
Epigallocatechin Gallate	Extract	mg/g	409	18	0.0	0.0	28	408	39	409	18
Epigallocatechin Gallate	Tea	mg/g	49.3	2.1	0.1	0.0	27	48.4	10.6	49.3	2.1
Gallocatechin	Extract	mg/g	21.3	1.6	-0.3	0.0	17	28.0	20.2	21.3	1.6
Gallocatechin	Tea	mg/g	2.28	1.04	-0.6	0.0	17	6.66	7.5	2.28	1.0
Gallocatechin Gallate	Extract	mg/g	37.8	1.9	-0.5	0.0	22	43.1	10.6	37.8	1.9
Gallocatechin Gallate	Tea	mg/g	0.939	0.199	-0.5	0.0	19	1.380	0.90	0.939	0.20
Total Catechins	Extract	mg/g	699	22	0.1	0.0	24	691	73	699	22
Total Catechins	Tea	mg/g	97.9	5.2	0.0	0.0	23	98.0	23.3	97.9	5.2

x_i Mean of reported values

s_i Standard deviation of reported values

Z_{comm} Z-score with respect to community consensus

Z_{NIST} Z-score with respect to NIST value

N   Number of quantitative values reported

x^* Robust mean of reported values

s^* Robust standard deviation

x_{NIST} NIST-assessed value

U_{95} ±95% confidence interval about the assessed value or standard deviation (s_{NIST})

Table 14. Data summary table for catechin in green tea.

	Lab	SRM 3255 Green Tea Extract (mg/g)					SRM 3254 Green Tea (mg/g)				
		A	B	C	Avg	SD	A	B	C	Avg	SD
Individual Results	NIST				8 88	0 90				0 958	0 389
	I901										
	I902	11 80	11 50	11 70	11 67	0 15	0 970	1 140	1 000	1 037	0 091
	I903	8 48	8 42	8 61	8 50	0 10	0 886	0 905	0 911	0 901	0 013
	I904										
	I905	10 30	10 20	10 20	10 23	0 06	1 700	1 700	1 700	1 700	0 000
	I906										
	I907										
	I909	4 71	5 10	5 22	5 01	0 27	0 201	0 205	0 188	0 198	0 009
	I911										
	I912	12 40	12 40	12 20	12 33	0 12	2 170	2 130	2 110	2 137	0 031
	I913	0 23	0 18	0 25	0 22	0 04					
	I914										
	I916										
	I918	10 09	10 45	10 43	10 32	0 20	1 159	1 037	0 968	1 055	0 097
	I921										
	I922										
	I923	8 73	8 81	8 77	8 77	0 04	0 780	0 870	0 770	0 807	0 055
	I926	12 20	12 30	12 60	12 37	0 21					
	I927	12 60	12 90	12 73	12 75	0 15	0 894	1 154	1 262	1 104	0 189
	I928										
	I930										
	I933	141 63	141 30	140 54	141 16	0 56	62 380	62 900	62 740	62 673	0 266
	I934	10 20	10 09	10 28	10 19	0 10	1 964	2 676	15 283	6 641	7 493
	I938										
	I939	10 23	10 12	10 07	10 14	0 08	2 816	2 772	2 863	2 817	0 045
	I940										
	I943	10 76	10 88	11 06	10 90	0 15	1 962	1 857	1 849	1 890	0 063
	I944										
	I946	10 70	10 59	10 61	10 63	0 06	0 907	0 907	0 914	0 909	0 004
	I947										
	I950										
	I952	9 22	9 32	9 72	9 42	0 26	1 240	1 180	1 190	1 203	0 032
	I953	111 63	113 17	111 07	111 96	1 09	59 450	60 930	62 330	60 903	1 440
	I954	4 83	4 73	4 75	4 77	0 05	0 617	0 602	0 624	0 614	0 011
	I956										
	I957										
	I958										
	I963	10 18	10 23	10 12	10 18	0 06	2 302	2 326	2 260	2 296	0 033
	I964	7 70	7 63	7 76	7 70	0 07	0 530	0 500	0 470	0 500	0 030
	I965										
	I966	8 40	7 80	8 00	8 07	0 31	1 000	1 200	1 100	1 100	0 100
	I968										
	I969										
	I970										
	I976										
	I979										
	I982	8 03	7 00	7 63	7 55	0 52		0 100		0 100	
	I984										
	I985	9 86	10 01	10 02	9 97	0 09	3 031	2 998	3 048	3 026	0 025
Community Results		Consensus Mean			9 87		Consensus Mean			1 602	
		Consensus Standard Deviation			2 68		Consensus Standard Deviation			1 229	
		Maximum			111 96		Maximum			60 903	
		Minimum			4 77		Minimum			0 100	
		N			11		N			10	

52

Table 15. Data summary table for epicatechin in green tea.

		Epicatechin									
		SRM 3255 Green Tea Extract (mg/g)					SRM 3254 Green Tea (mg/g)				
	Lab	A	B	C	Avg	SD	A	B	C	Avg	SD
Individual Results	NIST				45 8	6 5				8 53	1 52
	I901										
	I902	179 0	87 8	179 0	148 6	52 7	77 50	77 70	77 40	77 53	0 15
	I903	45 3	45 2	45 5	45 3	0 2	8 60	8 36	8 35	8 44	0 14
	I904										
	I905	47 4	47 3	47 2	47 3	0 1	8 40	8 50	8 30	8 40	0 10
	I906										
	I907										
	I909	34 3	37 4	41 5	37 7	3 6	7 72	7 40	7 42	7 51	0 18
	I911										
	I912	45 5	45 6	45 1	45 4	0 3	8 21	7 73	7 66	7 87	0 30
	I913	2 5	2 0	2 7	2 4	0 4	0 20	0 18	0 20	0 19	0 01
	I914										
	I916										
	I918	47 1	46 5	46 1	46 6	0 5	8 05	7 74	7 37	7 72	0 34
	I921										
	I922										
	I923	45 4	47 9	48 3	47 2	1 6	5 86	6 12	5 45	5 81	0 34
	I926	33 6	33 9	34 1	33 9	0 3					
	I927	48 3	48 8	48 3	48 5	0 3	6 01	8 24	8 11	7 45	1 25
	I928										
	I930										
	I933	46 2	45 6	45 5	45 8	0 4	6 43	6 69	6 54	6 55	0 13
	I934	43 5	44 3	43 8	43 9	0 4	6 24	6 07	6 79	6 37	0 37
	I938										
	I939	42 6	42 4	42 3	42 4	0 2	7 35	7 27	7 26	7 29	0 05
	I940										
	I943	43 2	43 8	43 8	43 6	0 4	7 65	7 50	7 50	7 55	0 09
	I944										
	I946	43 9	44 3	44 3	44 2	0 2	7 76	7 77	7 72	7 75	0 02
	I947										
	I950										
	I952	46 8	47 3	48 4	47 5	0 9	9 14	10 08	10 17	9 80	0 57
	I953	22 3	24 4	21 8	22 8	1 4	3 47	4 40	4 29	4 05	0 51
	I954	23 7	23 6	23 6	23 6	0 1	4 37	4 44	4 39	4 40	0 04
	I956										
	I957										
	I958										
	I963	47 6	48 5	47 1	47 7	0 7	11 29	9 34	11 08	10 57	1 07
	I964	40 1	39 7	40 4	40 1	0 4	7 17	6 84	6 71	6 91	0 24
	I965										
	I966	9 7	8 8	8 6	9 0	0 6	8 60	9 50	8 90	9 00	0 46
	I968	49 2	49 1	45 7	48 0	2 0	9 70	9 60	9 40	9 57	0 15
	I969										
	I970										
	I976										
	I979										
	I982	42 2	44 1	42 7	43 0	1 0	7 60	8 20	7 30	7 70	0 46
	I984										
	I985	43 0	42 9	43 2	43 1	0 1	7 57	7 58	7 41	7 52	0 10

Community Results	Consensus Mean	42 8	Consensus Mean	7 59
	Consensus Standard Deviation	6 4	Consensus Standard Deviation	1 74
	Maximum	48 0	Maximum	10 57
	Minimum	9 0	Minimum	4 05
	N	12	N	12

53

Table 16. Data summary table for epicatechin gallate in green tea.

		Epicatechin gallate									
		SRM 3255 Green Tea Extract (mg/g)					SRM 3254 Green Tea (mg/g)				
	Lab	A	B	C	Avg	SD	A	B	C	Avg	SD
Individual Results	NIST				97 2	7 6				12 0	1 1
	I901										
	I902	94 5	94 4	95 3	94 7	0 5	12 2	12 6	12 3	12 4	0 2
	I903	99 8	99 9	101 0	100 2	0 7	15 6	15 5	15 7	15 6	0 1
	I904										
	I905	98 9	99 5	98 5	99 0	0 5	14 1	14 2	14 2	14 2	0 1
	I906										
	I907										
	I909	111 3	107 0	116 1	111 4	4 5	12 3	12 8	12 6	12 6	0 2
	I911										
	I912	93 6	93 9	91 0	92 8	1 6	6 8	6 9	6 0	6 6	0 5
	I913	20 4	16 3	22 2	19 6	3 0	1 3	1 2	1 3	1 3	0 1
	I914										
	I916										
	I918	91 6	88 8	88 3	89 6	1 8	13 9	11 8	11 6	12 4	1 3
	I921										
	I922										
	I923	85 8	85 0	84 8	85 2	0 5	10 4	10 7	9 6	10 2	0 6
	I926	81 4	82 3	82 9	82 2	0 8					
	I927	97 5	104 0	105 9	102 5	4 4	6 9	10 1	10 9	9 3	2 1
	I928										
	I930										
	I933	108 5	106 2	107 0	107 2	1 1	11 0	11 0	11 0	11 0	0 0
	I934	95 9	97 1	96 7	96 6	0.6	12 8	13 3	13 5	13 2	0 4
	I938										
	I939	128 4	128 6	126 6	127 9	1 1	10 5	10 8	10 7	10 6	0 1
	I940										
	I943	90 6	94 1	93 8	92 8	1 9	12 8	12 3	12 2	12 4	0 3
	I944										
	I946	100 8	101 6	101 6	101 3	0 5	10 4	10 5	10 5	10 4	0 1
	I947										
	I950										
	I952	101 1	102 3	105 4	102 9	2 2	12 5	13 7	15 8	14 0	1 6
	I953	88 5	87 6	88 3	88 1	0 5	11 5	12 2	11 9	11 9	0 3
	I954	3 3	3 3	3 3	3 3	0 0	0 5	0 5	0 5	0 5	0 0
	I956										
	I957										
	I958										
	I963	81 6	81 7	81 0	81 4	0 4	9 9	9 8	9 7	9 8	0 1
	I964	100 0	99 1	101 0	100 0	1 0	12 4	11 9	11 7	12 0	0 4
	I965										
	I966	21 7	19 8	18 9	20 1	1 4	16 0	15 9	15 7	15 9	0 2
	I968	104 1	101 7	100 6	102 1	1 8	14 9	15 2	15 0	15 0	0 2
	I969										
	I970										
	I976										
	I979										
	I982	85 0	88 3	88 7	87 3	2 0	14 2	14 2	13 4	13 9	0 5
	I984										
	I985	121 4	121 4	122 0	121 6	0 4	11 0	10 5	10 5	10 7	0 3

Community Results											
	Consensus Mean				94 6		Consensus Mean				11 8
	Consensus Standard Deviation				13 8		Consensus Standard Deviation				2 8
	Maximum				127 9		Maximum				15 9
	Minimum				3 3		Minimum				0 5
	N				12		N				12

54

Table 17. Data summary table for epigallocatechin in green tea.

| | Lab | SRM 3255 Green Tea Extract (mg/g) | | | | | SRM 3254 Green Tea (mg/g) | | | | |
		A	B	C	Avg	SD	A	B	C	Avg	SD
	NIST				79 2	6 3				23 9	4 3
	I901										
	I902	88 5	87 8	89 1	88 5	0 7	19 8	20 0	20 3	20 0	0 3
	I903	77 3	76 6	77 7	77 2	0 6	27 9	26 5	27 3	27 2	0 7
	I904										
	I905	88 7	88 7	88 3	88 6	0 2	29 0	29 5	29 2	29 2	0 3
	I906										
	I907										
	I909	78 0	79 9	80 5	79 5	1 3	18 2	20 1	19 8	19 4	1 0
	I911										
	I912	30 1	30 1	29 3	29 8	0 5	4 7	4 6	4 6	4 6	0 1
	I913	1 3	1 0	1 4	1 3	0 2	0 2	0 2	0 2	0 2	0 0
	I914										
	I916										
	I918	68 2	70 0	71 3	69 8	1 6	24 6	22 6	21 3	22 8	1 7
	I921										
	I922										
	I923	72 7	72 9	70 7	72 1	1 3	18 0	19 1	17 0	18 0	1 0
	I926	64 9	64 7	66 0	65 2	0 7					
	I927	92 4	94 6	93 2	93 4	1 1	19 9	28 9	27 9	25 6	4 9
Individual Results	I928										
	I930										
	I933	75 5	75 3	75 8	75 5	0 3	20 2	21 3	20 8	20 8	0 6
	I934	79 5	80 9	80 5	80 3	0 7	23 3	25 8	26 3	25 1	1 6
	I938										
	I939	14 0	13 9	13 9	13 9	0 1	4 4	4 4	4 4	4 4	0 0
	I940										
	I943	68 4	70 5	70 4	69 8	1 2	24 6	23 3	22 9	23 6	0 9
	I944										
	I946	74 1	74 9	75 1	74 7	0 5	25 1	25 1	25 0	25 1	0 0
	I947										
	I950										
	I952	74 6	74 9	77 2	75 6	1 4	26 4	27 6	26 6	26 9	0 7
	I953	49 6	51 0	50 3	50 3	0 7	15 5	16 7	16 6	16 2	0 7
	I954	15 7	15 6	15 5	15 6	0 1	4 2	4 3	4 3	4 3	0 1
	I956										
	I957										
	I958										
	I963	123 1	123 5	126 4	124 3	1 8	36 0	36 2	35 4	35 9	0 4
	I964	89 0	88 4	89 8	89 1	0 7	27 2	25 7	25 2	26 0	1 0
	I965										
	I966	2 0	1 8	2 0	1 9	0 1	44 2	43 7	43 5	43 8	0 4
	I968	94 7	86 5	83 1	88 1	6 0	38 0	37 2	36 7	37 3	0 7
	I969										
	I970										
	I976										
	I979										
	I982										
	I984										
	I985	11 7	11 8	11 8	11 8	0 1	3 8	3 8	3 7	3 8	0 0
Community Results		Consensus Mean			63 8		Consensus Mean			20 8	
		Consensus Standard Deviation			33 3		Consensus Standard Deviation			12 5	
		Maximum			124 3		Maximum			43 8	
		Minimum			1 9		Minimum			3 8	
		N			11		N			11	

55

Table 18. Data summary table for epigallocatechin gallate in green tea.

	Lab	SRM 3255 Green Tea Extract (mg/g)					SRM 3254 Green Tea (mg/g)				
		A	B	C	Avg	SD	A	B	C	Avg	SD
Individual Results	NIST				409	18				49 3	2 1
	I901										
	I902	409	407	408	408	1	43 3	43 6	44 5	43 8	0 6
	I903	420	418	422	420	2	58 9	58 1	59 2	58 7	0 6
	I904										
	I905	458	462	457	459	3	59 6	60 1	60 5	60 1	0 5
	I906										
	I907										
	I909	358	375	394	376	18	65 2	63 5	64 4	64 4	0 8
	I911										
	I912	415	420	409	415	5	31 7	36 5	28 1	32 1	4 2
	I913	56	45	60	54	8	2 7	2 6	2 8	2 7	0 1
	I914										
	I916										
	I918	424	412	410	416	8	51 9	49 1	47 5	49 5	2 2
	I921										
	I922										
	I923	373	365	365	368	5	35 9	37 7	33 2	35 6	2 3
	I926	349	351	354	351	2					
	I927	402	411	408	407	4	34 2	50 1	50 1	44 8	9 2
	I928	428	426	432	429	3	57 3	54 8	56 3	56 1	1 3
	I930										
	I933	404	402	400	402	2	37 1	38 3	37 8	37 7	0 6
	I934	409	412	411	411	2	52 5	52 6	53 6	52 9	0 6
	I938										
	I939	462	464	460	462	2	39 5	40 6	40 2	40 1	0 6
	I940	437	429	434	434	4	45 5	45 1	46 5	45 7	0 7
	I943	395	400	399	398	2	49 0	47 5	46 8	47 8	1 1
	I944										
	I946	402	407	407	405	3	41 9	42 1	42 2	42 1	0 2
	I947										
	I950										
	I952	404	409	421	411	9	48 7	53 1	58 3	53 4	4 8
	I953	356	367	358	360	6	38 0	42 2	42 1	40 8	2 4
	I954	438	437	434	436	2	59 5	60 6	59 5	59 9	0 6
	I956										
	I957										
	I958										
	I963	408	403	399	403	5	45 6	45 5	45 5	45 5	0 1
	I964	436	433	441	437	4	49 2	46 8	46 1	47 4	1 6
	I965										
	I966	89	82	78	83	6	54 6	53 4	51 7	53 2	1 5
	I968	406	406	399	404	4	60 2	59 6	58 5	59 4	0 9
	I969										
	I970										
	I976										
	I979										
	I982	449	465	463	459	9	66 3	67 5	62 5	65 4	2 6
	I984										
	I985	463	464	466	464	1	40 3	39 3	38 8	39 4	0 8
Community Results		Consensus Mean			409		Consensus Mean			48 3	
		Consensus Standard Deviation			39		Consensus Standard Deviation			11 0	
		Maximum			464		Maximum			65 4	
		Minimum			83		Minimum			39 4	
		N			13		N			13	

Table 19. Data summary table for gallocatechin in green tea.

	Lab	SRM 3255 Green Tea Extract (mg/g)					SRM 3254 Green Tea (mg/g)				
		A	B	C	Avg	SD	A	B	C	Avg	SD
Individual Results	NIST				21 3	1 6				2 28	1 04
	I901										
	I902										
	I903	12 9	12 6	12 5	12 7	0 2	6 57	6 84	7 15	6 85	0 29
	I904										
	I905	24 3	24 2	24 3	24 3	0 1	4 20	4 10	4 20	4 17	0 06
	I906										
	I907										
	I909	14 6	17 1	18 8	16 8	2 1	1 13	1 12	1 07	1 10	0 03
	I911										
	I912	99 9	100 1	100 4	100 1	0 3	31 15	29 03	29 11	29 76	1 20
	I913	0 1		0 1	0 1	0 0	0 02	0 02		0 02	0 00
	I914										
	I916										
	I918	5 9	5 7	5 8	5 8	0 1	0 30	0 26	0 29	0 28	0 02
	I921										
	I922										
	I923	18 1	19 7	19 3	19 0	0 8	1 78	2 17	1 82	1 92	0 21
	I926										
	I927	28 2	29 3	29 1	28 9	0 6	3 14	4 00	4 52	3 89	0 70
	I928										
	I930										
	I933	113 7	113 2	115 6	114 2	1 2	17 48	17 07	17 61	17 39	0 28
	I934	22 6	23 3	22 5	22 8	0 4	2 41	2 43	2 61	2 48	0 11
	I938										
	I939										
	I940										
	I943										
	I944										
	I946	26 1	26 4	26 4	26 3	0 2	4 08	4 09	4 07	4 08	0 01
	I947										
	I950										
	I952	22 8	22 9	23 6	23 1	0 5	3 72	5 40	3 66	4 26	0 99
	I953	15 4	24 5	22 2	20 7	4 8	6 13	5 86	6 34	6 11	0 24
	I954	131 0	131 0	129 0	130 3	1 2	19 40	19 70	19 30	19 47	0 21
	I956										
	I957										
	I958										
	I963	28 7	28 9	28 6	28 7	0 2	4 52	4 51	4 46	4 50	0 03
	I964										
	I965										
	I966										
	I968	47 9	55 5	45 9	49 8	5 1	27 78	27 36	29 06	28 07	0 89
	I969										
	I970										
	I976										
	I979										
	I982	22 3	23 4	22 9	22 9	0 6	3 30	2 60	2 21	2 70	0 55
	I984										
	I985										
Community Results		Consensus Mean			26 6		Consensus Mean			5 42	
		Consensus Standard Deviation			16 9		Consensus Standard Deviation			4 99	
		Maximum			130 3		Maximum			28 07	
		Minimum			20 7		Minimum			2 70	
		N			7		N			7	

57

Table 20. Data summary table for gallocatechin gallate in green tea.

		Gallocatechin gallate									
		SRM 3255 Green Tea Extract (mg/g)					SRM 3254 Green Tea (mg/g)				
	Lab	A	B	C	Avg	SD	A	B	C	Avg	SD
Individual Results	NIST				37 8	1 9				0 939	0 199
	I901										
	I902	44 8	45 2	46 2	45 4	0 7	1 910	1 840	2 090	1 947	0 129
	I903	42 8	42 8	42 9	42 8	0 1	1 260	1 190	1 160	1 203	0 051
	I904										
	I905	1 4	1 4	1 5	1 4	0 1	0 100	0 100	0 100	0 100	0 000
	I906										
	I907										
	I909	37 7	37 2	49 1	41 3	6 7	1 279	1 303	1 282	1 288	0 013
	I911										
	I912	36 0	36 5	35 2	35 9	0 7					
	I913	7 0	5 5	7 5	6 7	1 0	0 060	0 060	0 060	0 060	0 000
	I914										
	I916										
	I918	43 0	42 2	42 1	42 4	0 5	1 129	1 163	1 281	1 191	0 080
	I921										
	I922										
	I923	41 6	41 2	41 1	41 3	0 3	1 120	1 140	1 030	1 097	0 059
	I926	42 2	41 9	42 5	42 2	0 3					
	I927	101 5	106 1	105 2	104 3	2 5	2 700	2 473	4 277	3 150	0 983
	I928										
	I930										
	I933	61 2	52 3	52 6	55 4	5 1	1 310	1 380	1 360	1 350	0 036
	I934	53 4	54 2	54 1	53 9	0 4	1 205	1 147	1 266	1 206	0 060
	I938										
	I939	39 2	38 6	38 4	38 7	0 4	0 877	0 990	1 044	0 971	0 085
	I940										
	I943										
	I944										
	I946	55 3	55 3	55 6	55 4	0 2	1 776	1 795	1 785	1 786	0 010
	I947										
	I950										
	I952	42 4	42 8	43 9	43 0	0 8	2 060	2 020	2 510	2 197	0 272
	I953	15 8	16 9	15 4	16 0	0 8	0 414	0 041	0 381	0 279	0 206
	I954	47 2	47 1	46 6	47 0	0 3	1 420	1 440	1 370	1 410	0 036
	I956										
	I957										
	I958										
	I963	57 8	55 3	55 0	56 0	1 5	2 616	2 430	2 560	2 535	0 095
	I964	41 9	41 7	42 6	42 1	0 5	1 040	1 000	0 950	0 997	0 045
	I965										
	I966										
	I968	52 7	53 0	49 0	51 6	2 2	4 266	4 268	3 085	3 873	0 682
	I969										
	I970										
	I976										
	I979										
	I982	37 1	37 8	38 4	37 7	0 7		0 800		0 800	
	I984										
	I985	35 4	35 7	35 9	35 7	0 2	1 314	1 234	1 116	1 221	0 100

Community Results	Consensus Mean	43 1	Consensus Mean	1 342
	Consensus Standard Deviation	10 6	Consensus Standard Deviation	0 847
	Maximum	56 0	Maximum	3 873
	Minimum	16 0	Minimum	0 279
	N	10	N	9

Table 21. Data summary table for total catechins in green tea.

	Lab	SRM 3255 Green Tea Extract (mg/g)					SRM 3254 Green Tea (mg/g)				
		A	B	C	Avg	SD	A	B	C	Avg	SD
Individual Results	NIST				699	22				97 9	5 2
	I901										
	I902	828	825	829	827	2	155 7	156 9	157 6	156 7	1 0
	I903	712	709	715	712	3	120 0	118 0	120 0	119 3	1 2
	I904										
	I905	734	739	732	735	3	117 4	118 5	118 5	118 1	0 6
	I906										
	I907										
	I909	638	659	705	668	34	106 0	106 4	106 8	106 4	0 4
	I911										
	I912	733	739	722	731	8	84 8	86 8	77 6	83 1	4 8
	I913	88	70	94	84	13	4 5	4 2	4 6	4 4	0 2
	I914										
	I916										
	I918	690	676	674	680	9	101 0	93 6	90 3	95 0	5 5
	I921										
	I922										
	I923	645	640	638	641	4	73 9	77 8	68 8	73 5	4 5
	I926	583	586	592	587	4					
	I927	783	807	802	797	13	73 7	104 9	107 2	95 3	18 7
	I928										
	I930										
	I933	951	936	937	941	8	156 0	158 7	157 8	157 5	1 4
	I934	714	722	72	503	373	100 5	104 0	105 6	103 3	2 6
	I938										
	I939	696	697	691	695	3	65 5	66 8	66 4	66 2	0 7
	I940										
	I943	608	619	618	615	6	96 0	92 4	91 2	93 2	2 5
	I944										
	I946	713	720	721	718	4	91 9	92 3	92 2	92 1	0 2
	I947										
	I950										
	I952	701	708	729	713	15	103 7	113 1	118 2	111 7	7 3
	I953	663	687	670	673	13	135 2	143 2	144 6	141 0	5 1
	I954	664	662	657	661	4	90 0	91 6	90 0	90 5	0 9
	I956										
	I957										
	I958										
	I963	753	751	747	750	3	112 2	110 1	109 9	110 7	1 3
	I964	715	710	723	716	7	97 5	92 8	91 1	93 8	3 3
	I965										
	I966										
	I968	755	752	723	743	17	154 9	153 3	151 8	153 3	1 5
	I969										
	I970										
	I976										
	I979										
	I982	649	671	669	663	12	91 4	93 4	85 4	90 1	4 2
	I984										
	I985	684	685	689	686	2	66 9	65 4	64 6	65 7	1 2

Community Results	Consensus Mean	694	Consensus Mean	102 0
	Consensus Standard Deviation	73	Consensus Standard Deviation	29 7
	Maximum	750	Maximum	153 3
	Minimum	615	Minimum	65 7
	N	11	N	11

59

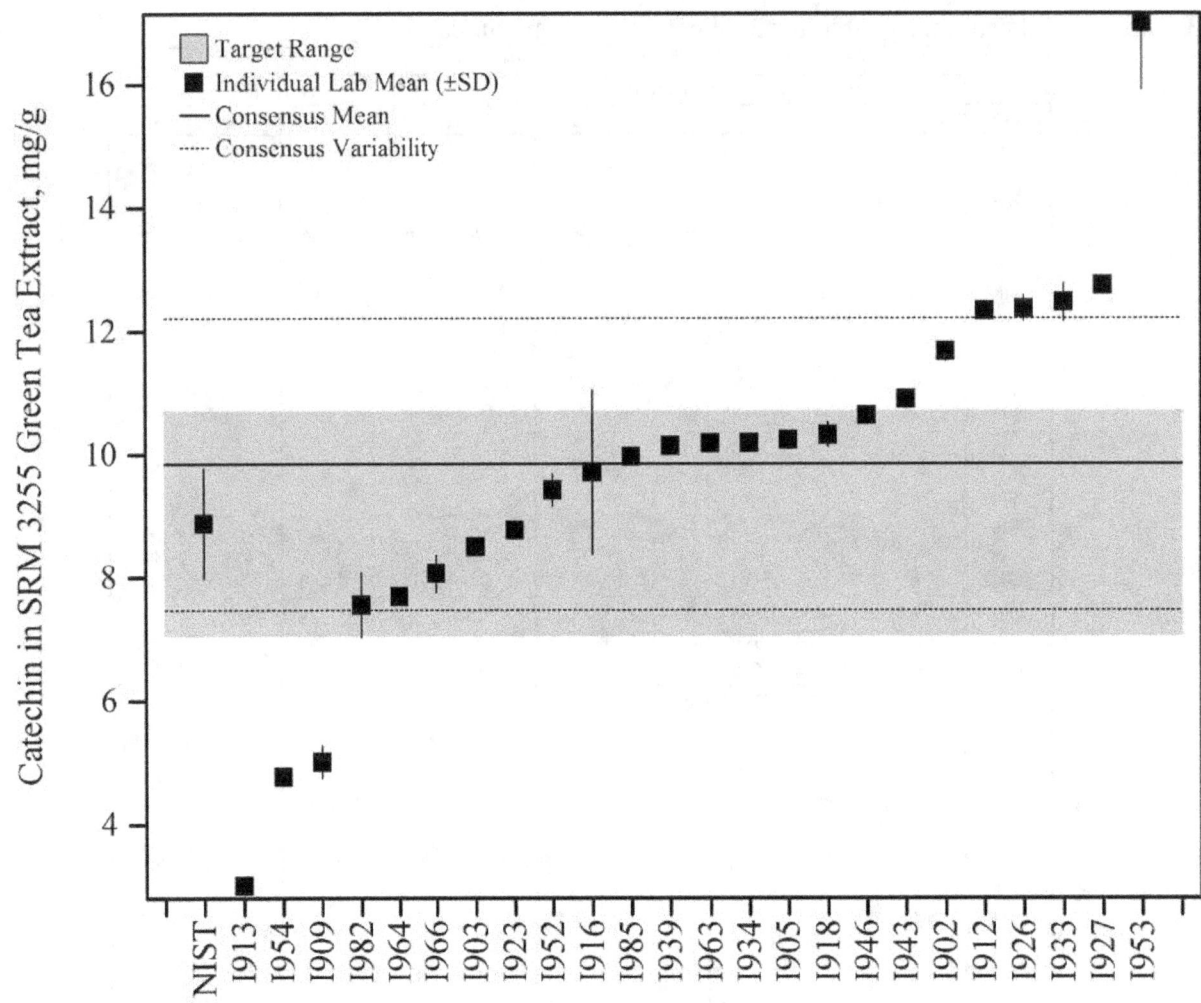

Figure 22. Catechin in SRM 3255 *Camellia sinensis* (Green Tea) Extract (data summary view). In this view, individual laboratory data are plotted with the individual laboratory standard deviation (error bars). The black solid line represents the consensus mean, and the black dotted lines represent the consensus variability calculated as one standard deviation about the consensus mean. The gray shaded region represents the target zone for "acceptable" performance, which encompasses the NIST certified value bounded by twice its uncertainty (U_{95}).

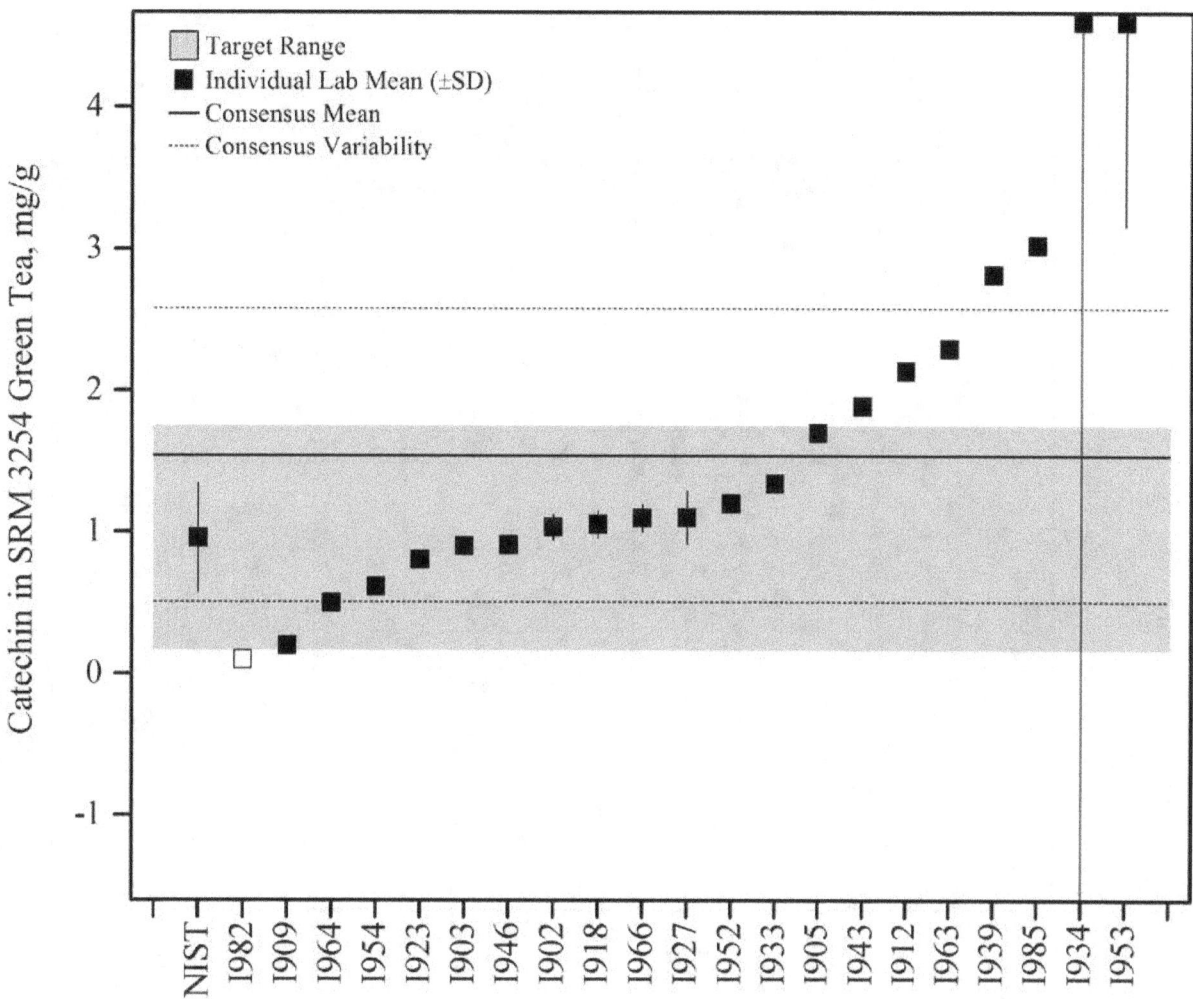

Figure 23. Catechin in SRM 3254 *Camellia sinensis* (Green Tea) (data summary view). In this view, individual laboratory data are plotted with the individual laboratory standard deviation (error bars). Data points that are unfilled represent laboratories that only reported a single value for that analyte and therefore were not included in the consensus mean. The black solid line represents the consensus mean, and the black dotted lines represent the consensus variability calculated as one standard deviation about the consensus mean. The gray shaded region represents the target zone for "acceptable" performance, which encompasses the NIST certified value bounded by twice its uncertainty (U_{95}).

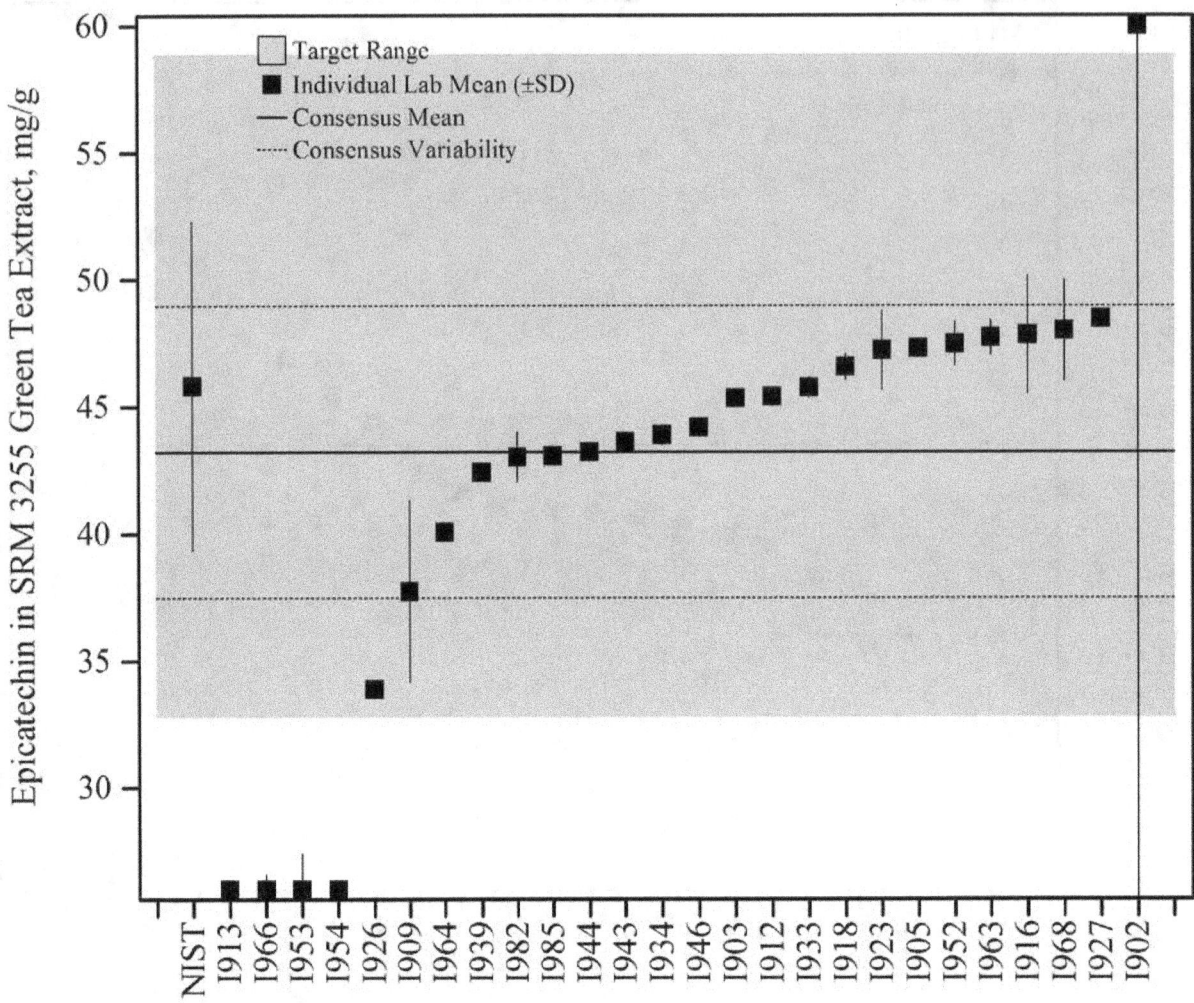

Figure 24. Epicatechin in SRM 3255 *Camellia sinensis* (Green Tea) Extract (data summary view). In this view, individual laboratory data are plotted with the individual laboratory standard deviation (error bars). The black solid line represents the consensus mean, and the black dotted lines represent the consensus variability calculated as one standard deviation about the consensus mean. The gray shaded region represents the target zone for "acceptable" performance, which encompasses the NIST certified value bounded by twice its uncertainty (U_{95}).

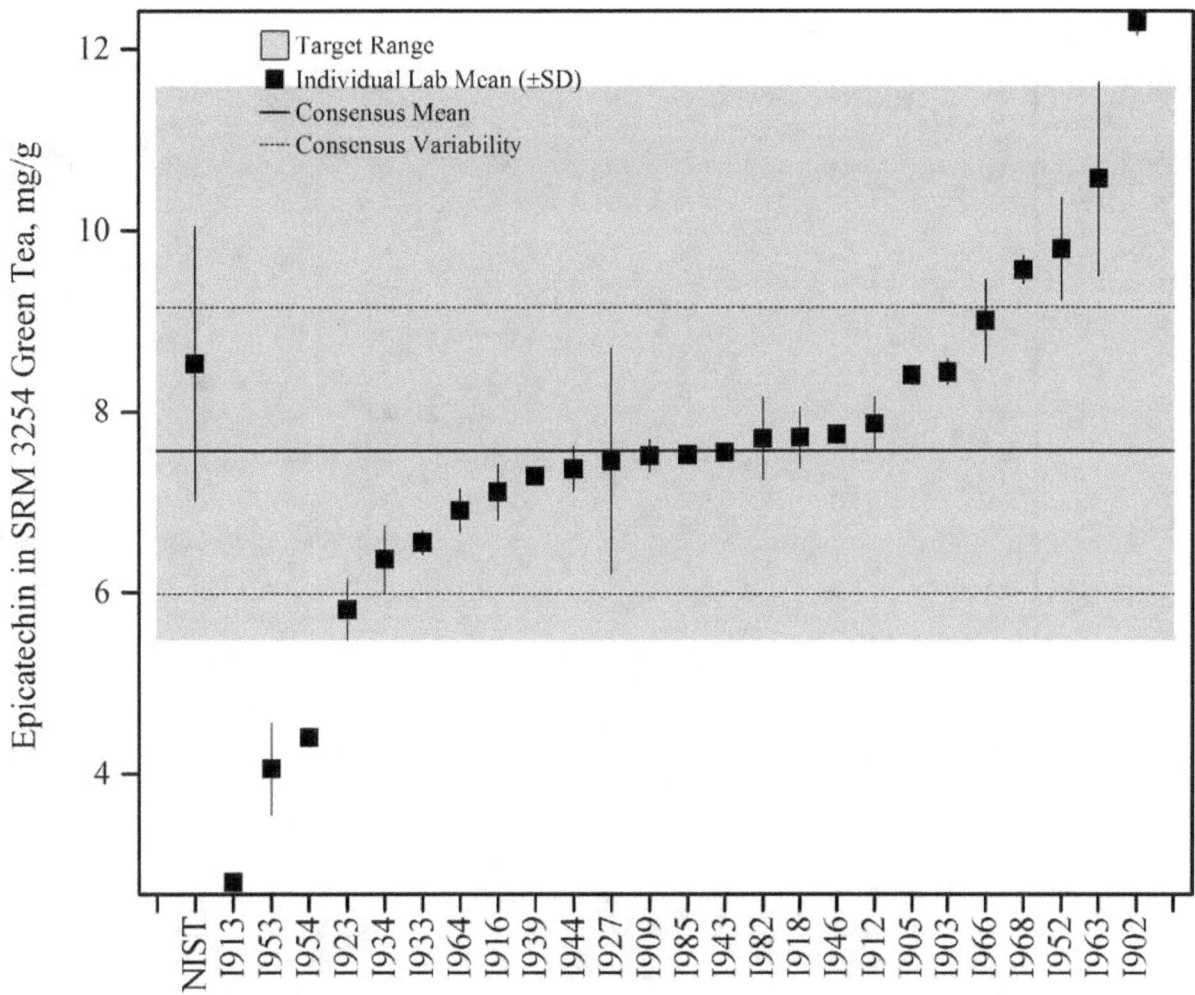

Figure 25. Epicatechin in SRM 3254 *Camellia sinensis* (Green Tea) (data summary view). In this view, individual laboratory data are plotted with the individual laboratory standard deviation (error bars). The black solid line represents the consensus mean, and the black dotted lines represent the consensus variability calculated as one standard deviation about the consensus mean. The gray shaded region represents the target zone for "acceptable" performance, which encompasses the NIST certified value bounded by twice its uncertainty (U_{95}).

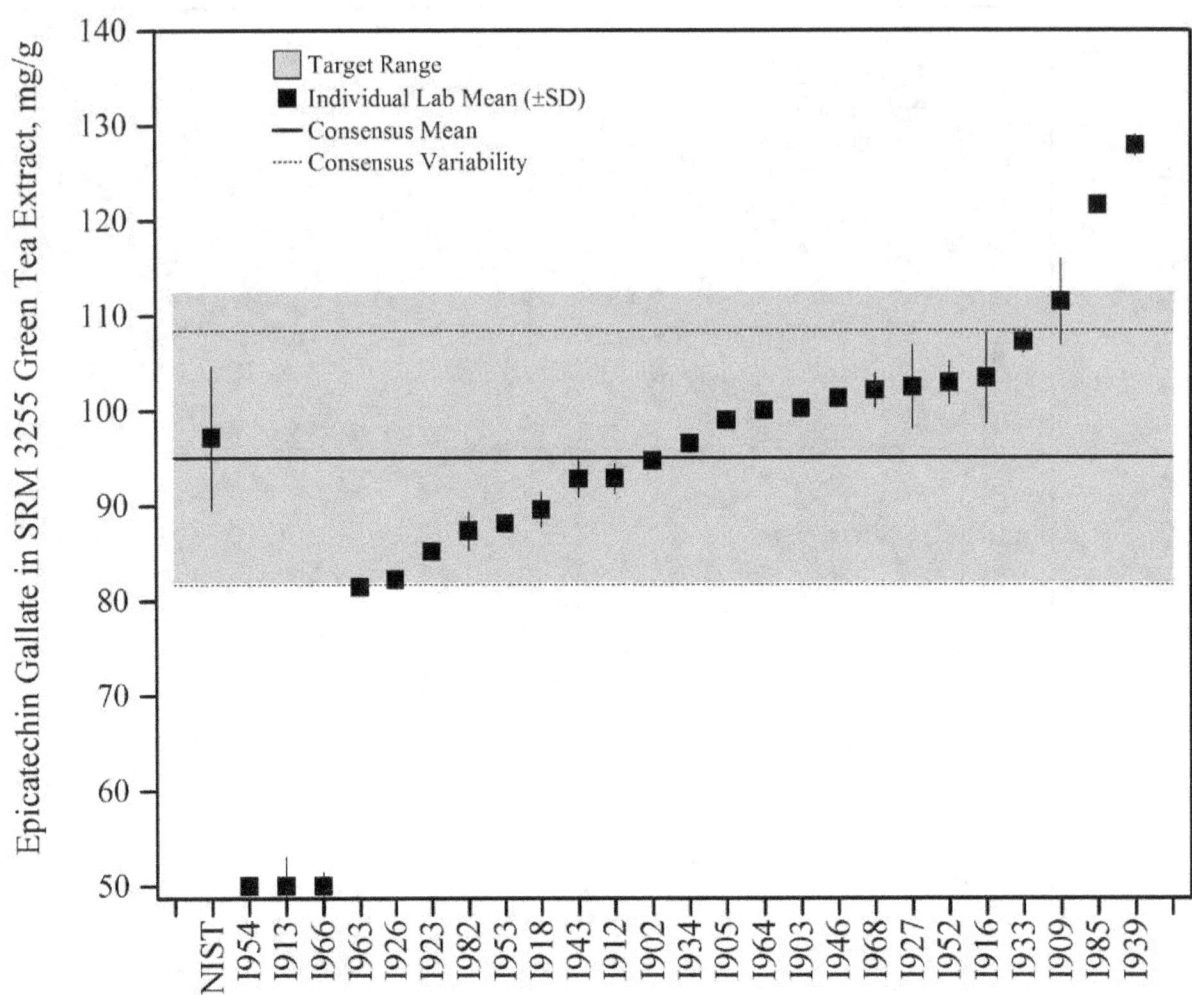

Figure 26. Epicatechin gallate in SRM 3255 *Camellia sinensis* (Green Tea) Extract (data summary view). In this view, individual laboratory data are plotted with the individual laboratory standard deviation (error bars). The black solid line represents the consensus mean, and the black dotted lines represent the consensus variability calculated as one standard deviation about the consensus mean. The gray shaded region represents the target zone for "acceptable" performance, which encompasses the NIST certified value bounded by twice its uncertainty (U_{95}).

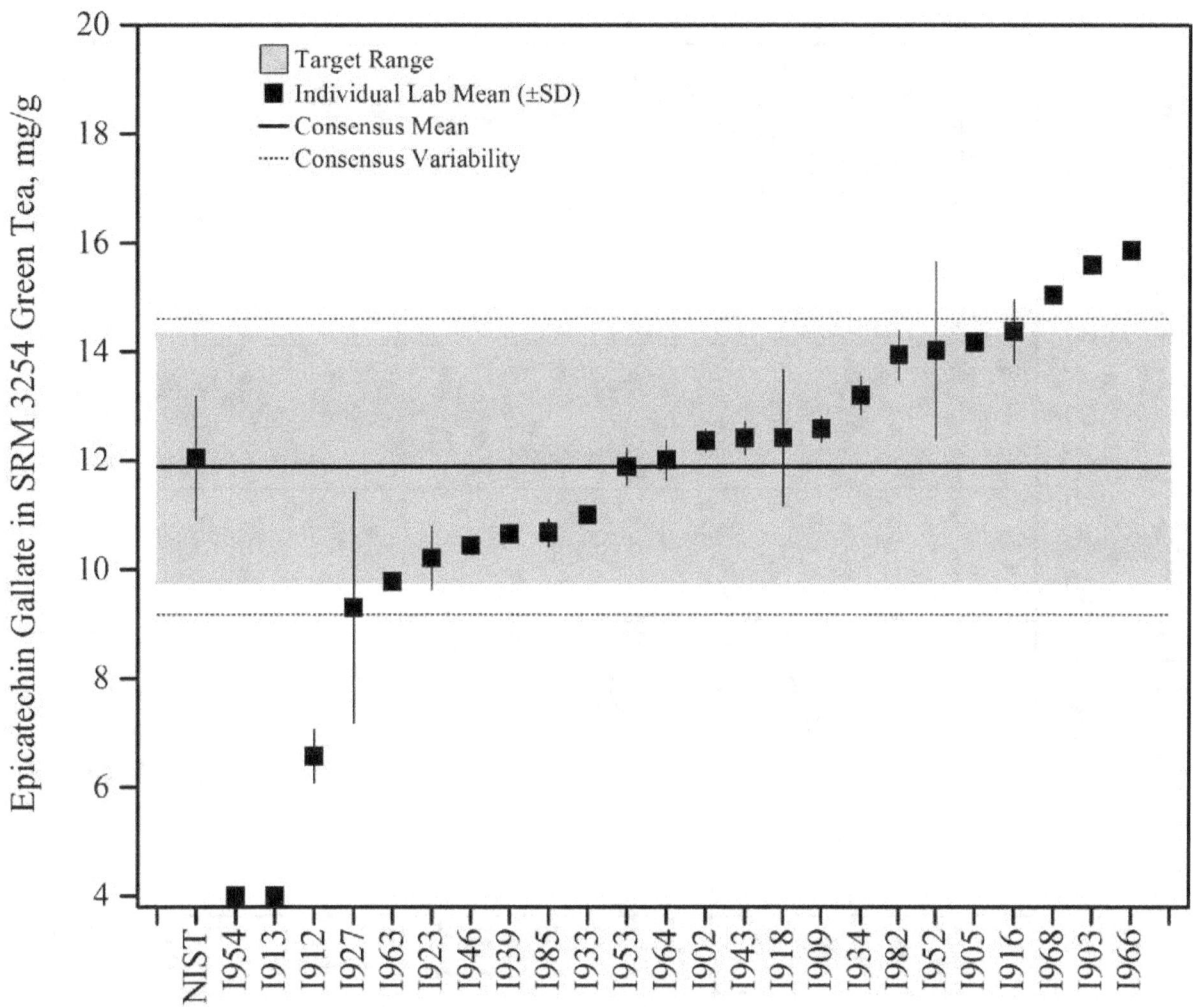

Figure 27. Epicatechin gallate in SRM 3254 *Camellia sinensis* (Green Tea) (data summary view). In this view, individual laboratory data are plotted with the individual laboratory standard deviation (error bars). The black solid line represents the consensus mean, and the black dotted lines represent the consensus variability calculated as one standard deviation about the consensus mean. The gray shaded region represents the target zone for "acceptable" performance, which encompasses the NIST certified value bounded by twice its uncertainty (U_{95}).

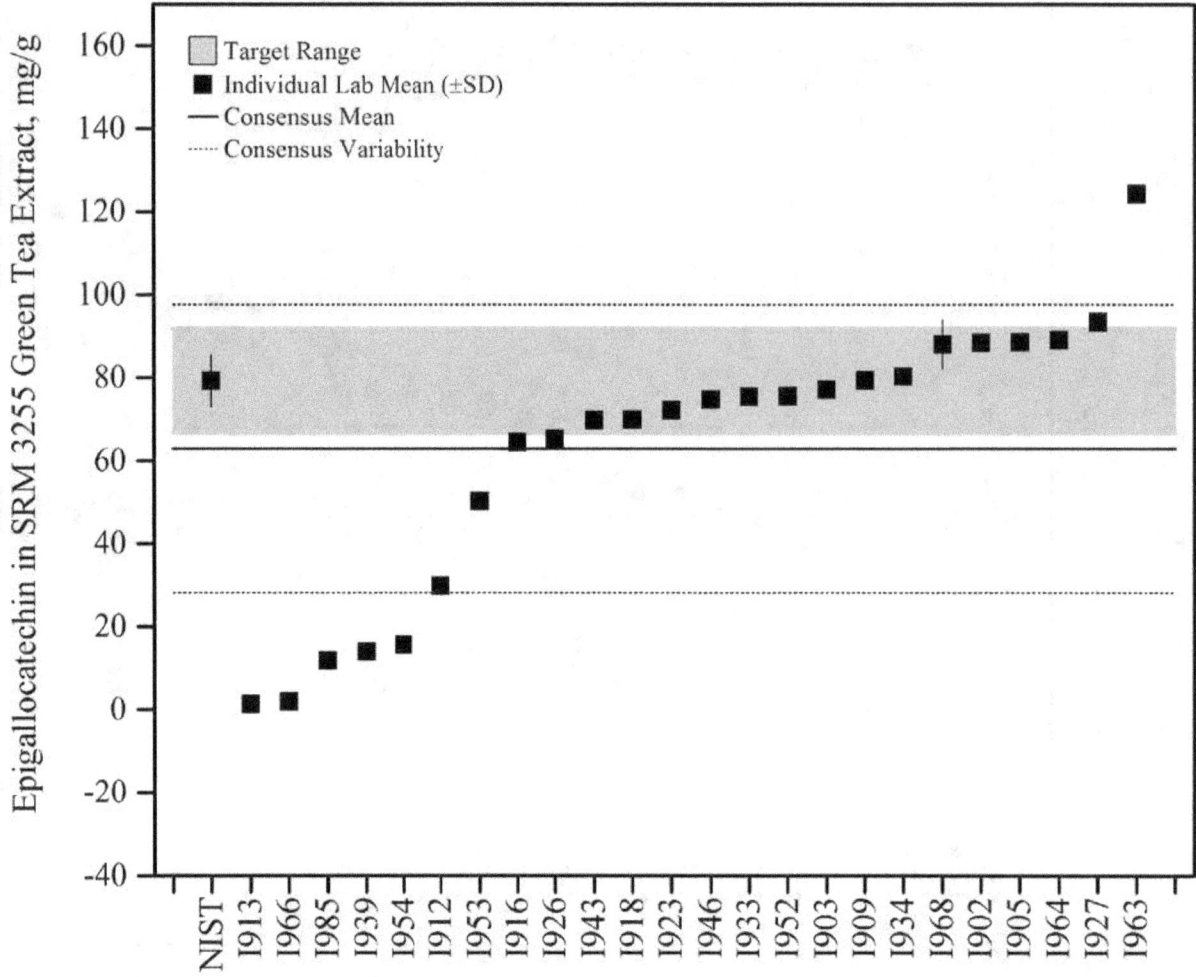

Figure 28. Epigallocatechin in SRM 3255 *Camellia sinensis* (Green Tea) Extract (data summary view). In this view, individual laboratory data are plotted with the individual laboratory standard deviation (error bars). The black solid line represents the consensus mean, and the black dotted lines represent the consensus variability calculated as one standard deviation about the consensus mean. The gray shaded region represents the target zone for "acceptable" performance, which encompasses the NIST certified value bounded by twice its uncertainty (U_{95}).

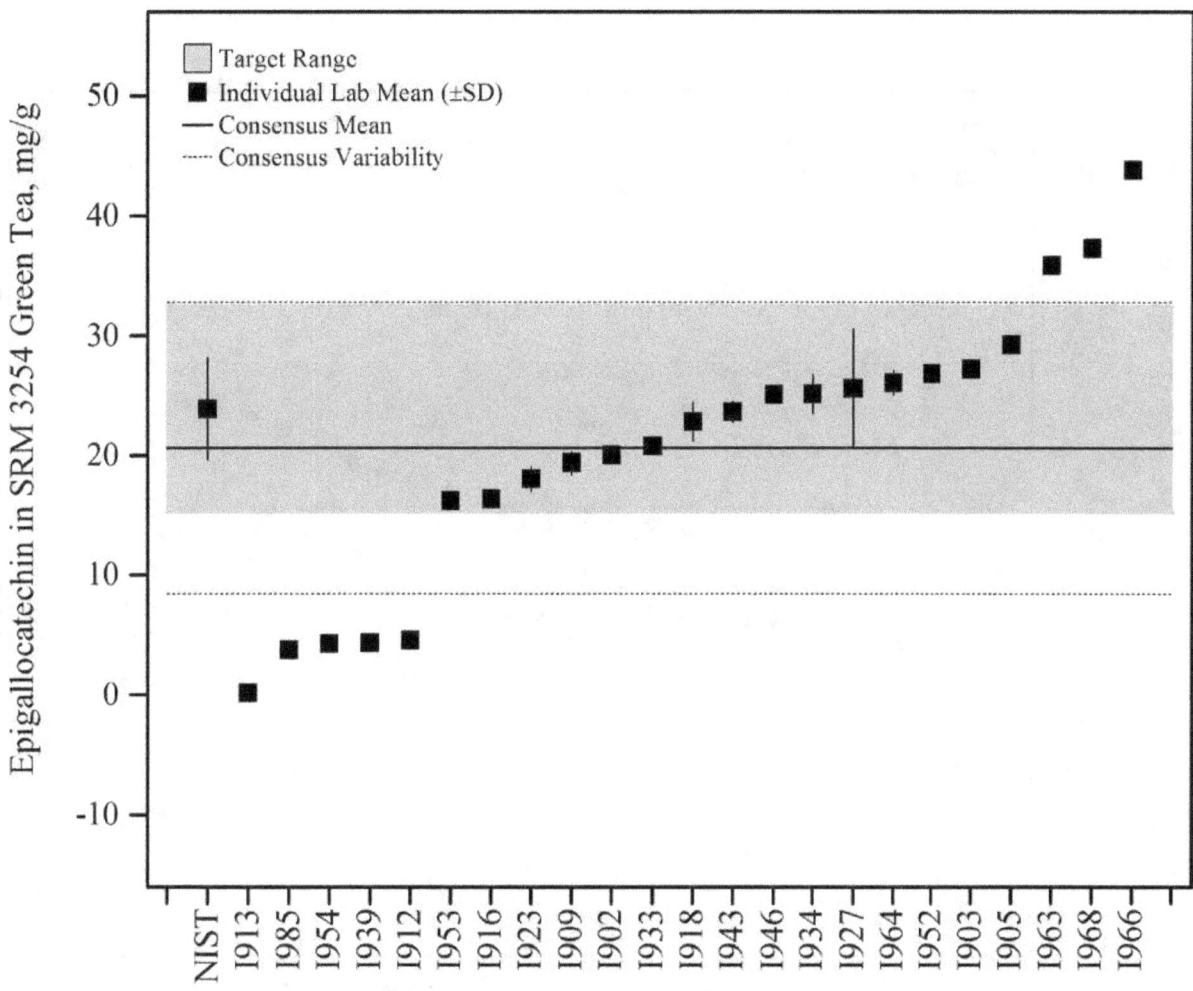

Figure 29. Epigallocatechin in SRM 3254 *Camellia sinensis* (Green Tea) (data summary view). In this view, individual laboratory data are plotted with the individual laboratory standard deviation (error bars). The black solid line represents the consensus mean, and the black dotted lines represent the consensus variability calculated as one standard deviation about the consensus mean. The gray shaded region represents the target zone for "acceptable" performance, which encompasses the NIST certified value bounded by twice its uncertainty (U_{95}).

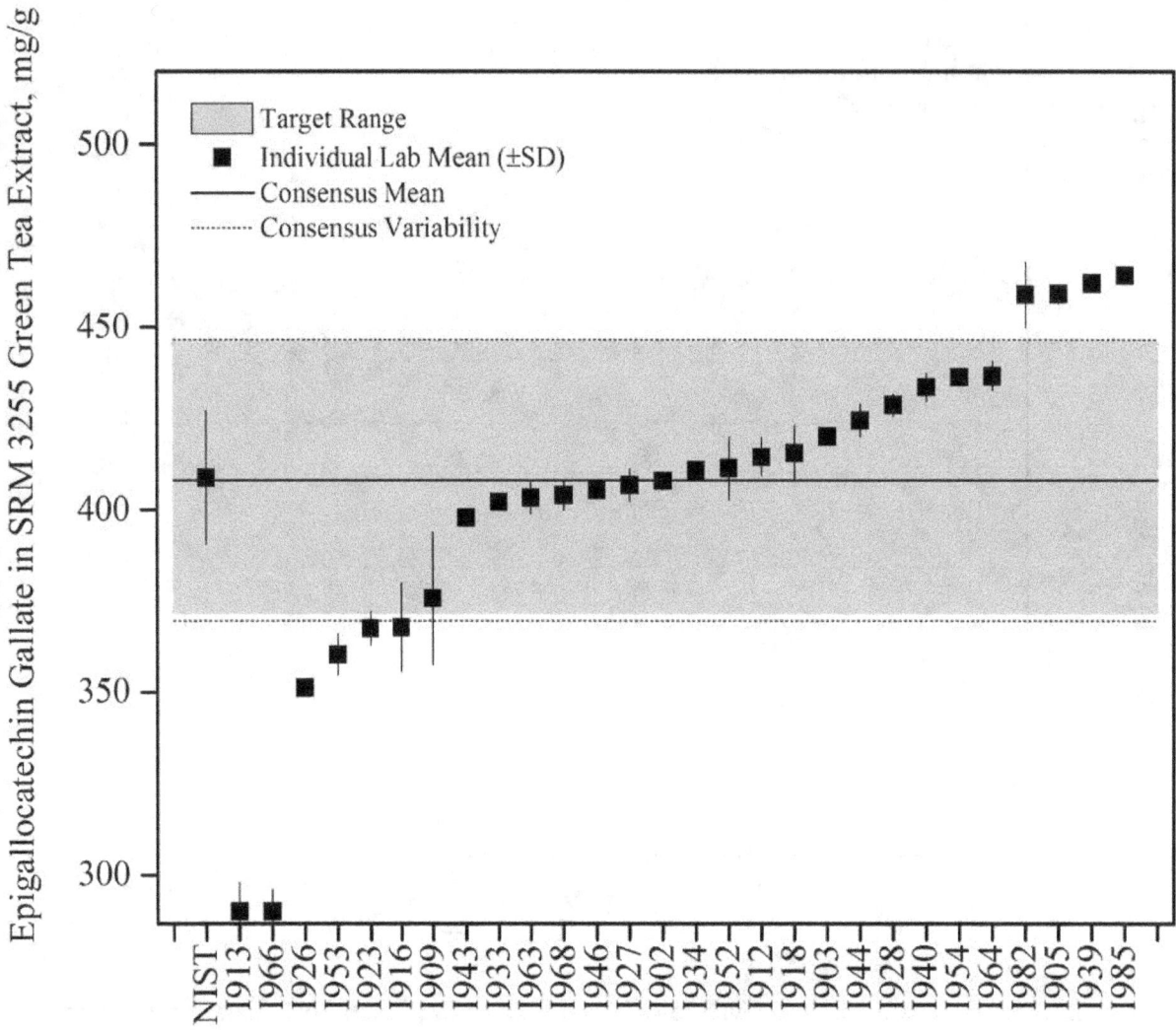

Figure 30. Epigallocatechin gallate in SRM 3255 *Camellia sinensis* (Green Tea) Extract (data summary view). In this view, individual laboratory data are plotted with the individual laboratory standard deviation (error bars). The black solid line represents the consensus mean, and the black dotted lines represent the consensus variability calculated as one standard deviation about the consensus mean. The gray shaded region represents the target zone for "acceptable" performance, which encompasses the NIST certified value bounded by twice its uncertainty (U_{95}).

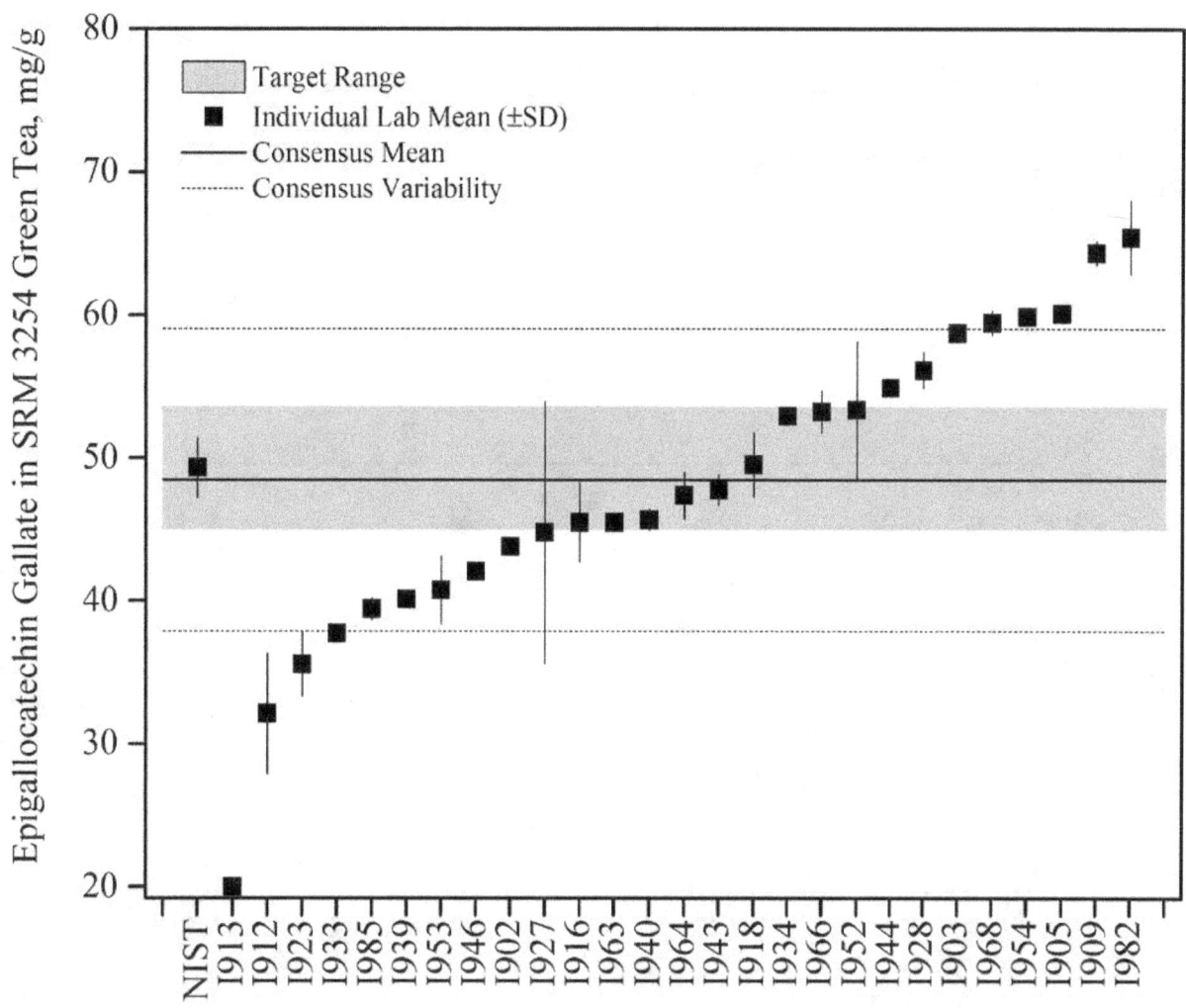

Figure 31. Epigallocatechin gallate in SRM 3254 *Camellia sinensis* (Green Tea) (data summary view). In this view, individual laboratory data are plotted with the individual laboratory standard deviation (error bars). The black solid line represents the consensus mean, and the black dotted lines represent the consensus variability calculated as one standard deviation about the consensus mean. The gray shaded region represents the target zone for "acceptable" performance, which encompasses the NIST certified value bounded by twice its uncertainty (U_{95}).

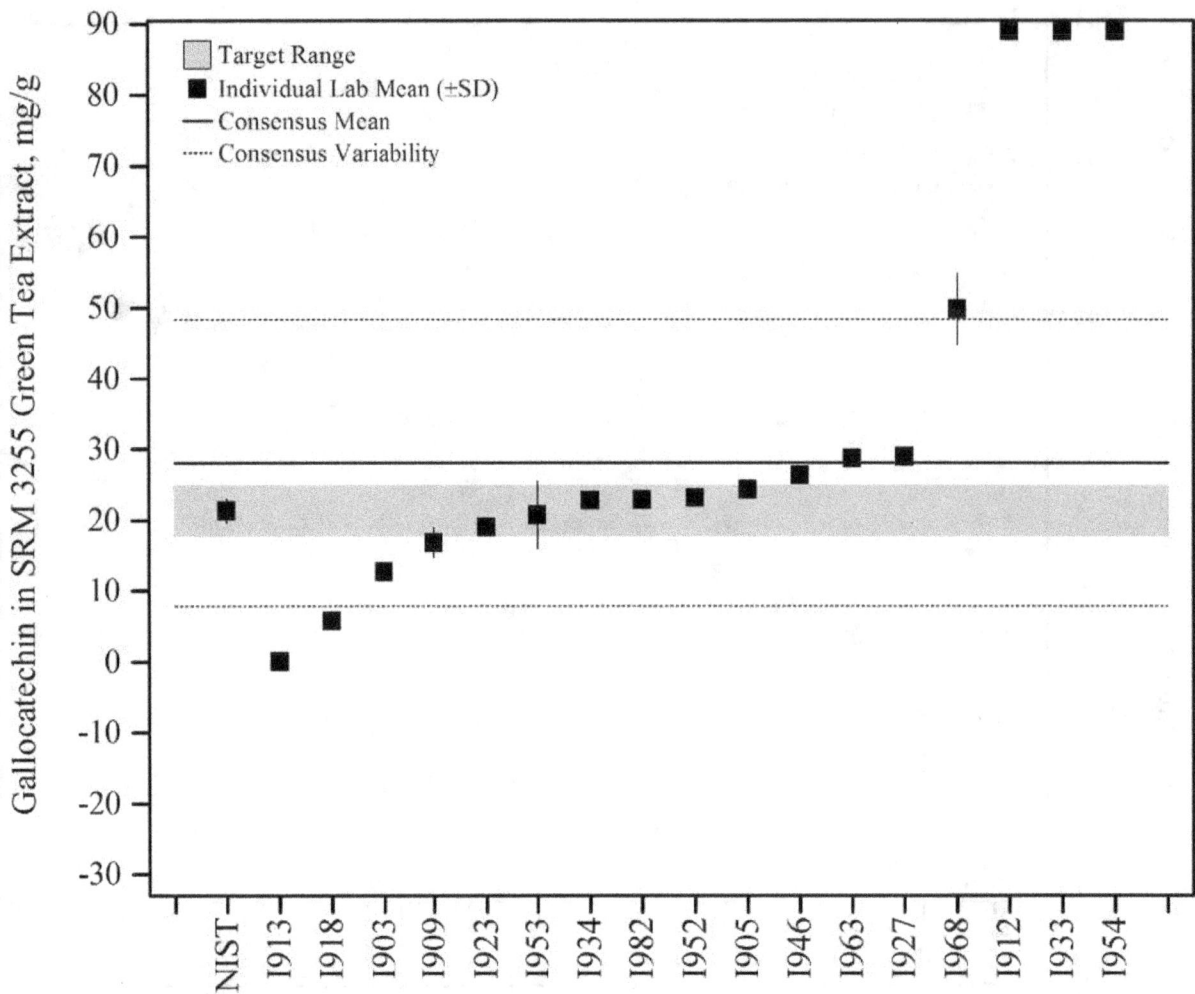

Figure 32. Gallocatechin in SRM 3255 *Camellia sinensis* (Green Tea) Extract (data summary view). In this view, individual laboratory data are plotted with the individual laboratory standard deviation (error bars). The black solid line represents the consensus mean, and the black dotted lines represent the consensus variability calculated as one standard deviation about the consensus mean. The gray shaded region represents the target zone for "acceptable" performance, which encompasses the NIST certified value bounded by twice its uncertainty (U_{95}).

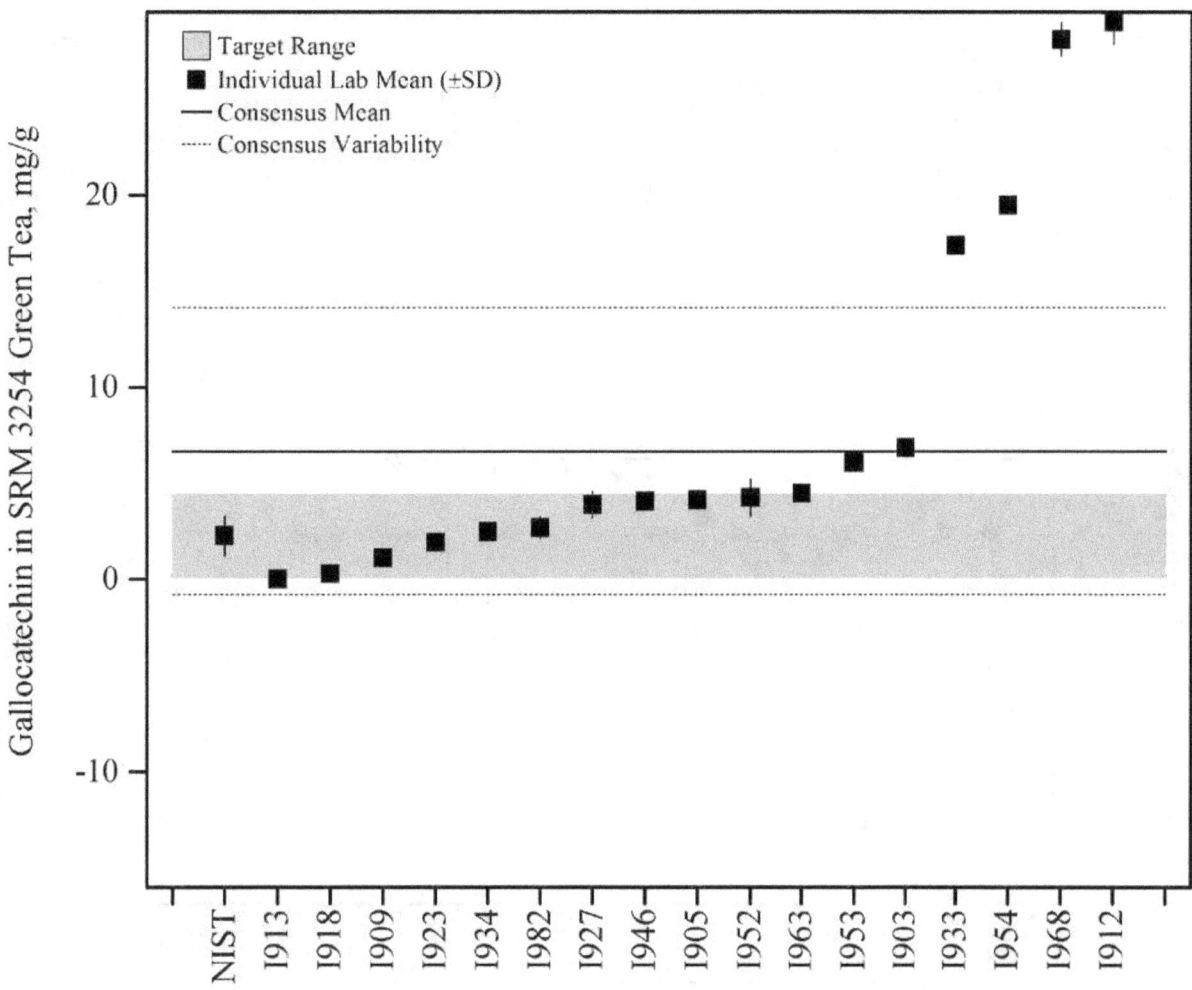

Figure 33. Gallocatechin in SRM 3254 *Camellia sinensis* (Green Tea) (data summary view). In this view, individual laboratory data are plotted with the individual laboratory standard deviation (error bars). The black solid line represents the consensus mean, and the black dotted lines represent the consensus variability calculated as one standard deviation about the consensus mean. The gray shaded region represents the target zone for "acceptable" performance, which encompasses the NIST certified value bounded by twice its uncertainty (U_{95}).

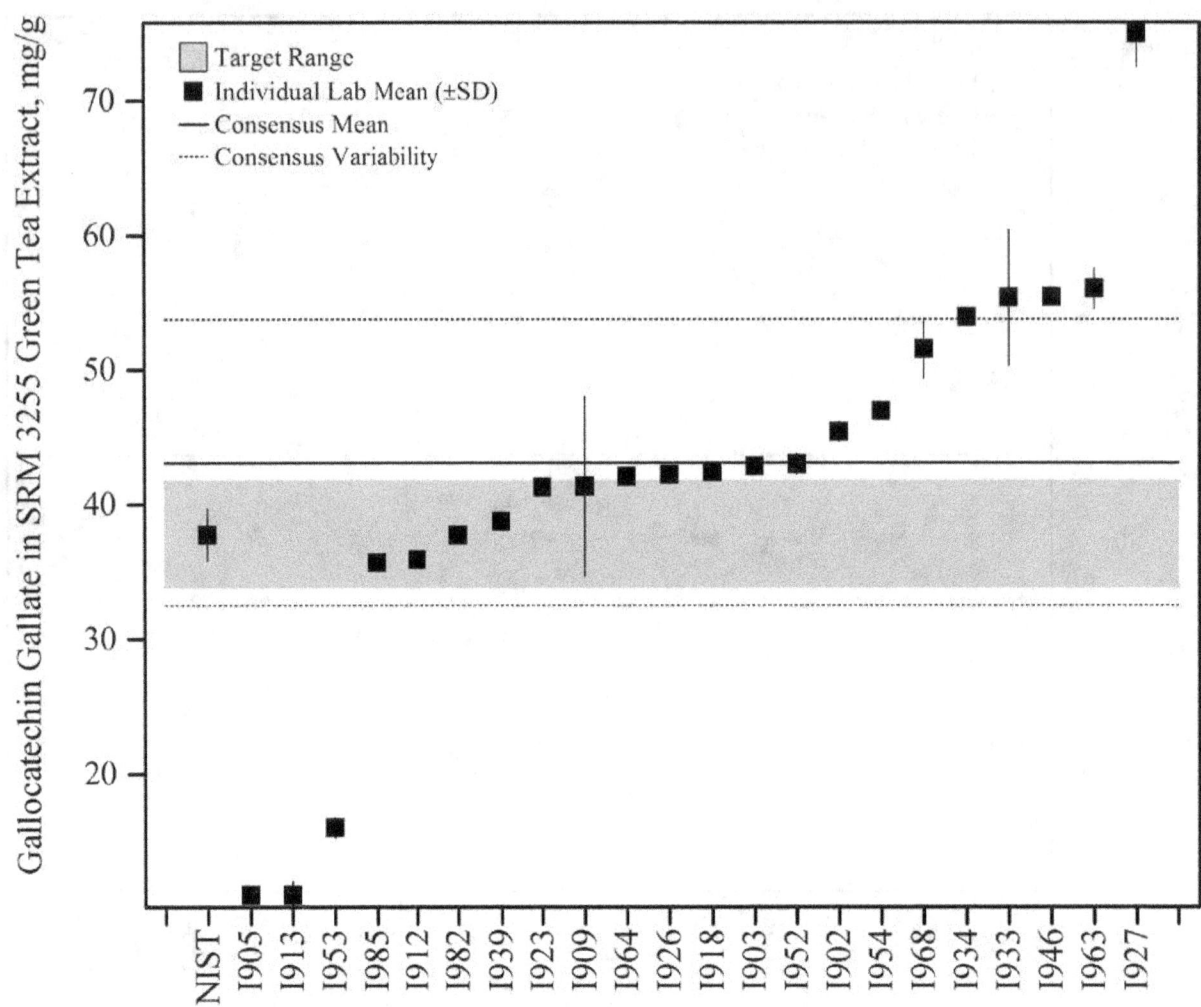

Figure 34. Gallocatechin gallate in SRM 3255 *Camellia sinensis* (Green Tea) Extract (data summary view). In this view, individual laboratory data are plotted with the individual laboratory standard deviation (error bars). The black solid line represents the consensus mean, and the black dotted lines represent the consensus variability calculated as one standard deviation about the consensus mean. The gray shaded region represents the target zone for "acceptable" performance, which encompasses the NIST certified value bounded by twice its uncertainty (U_{95}).

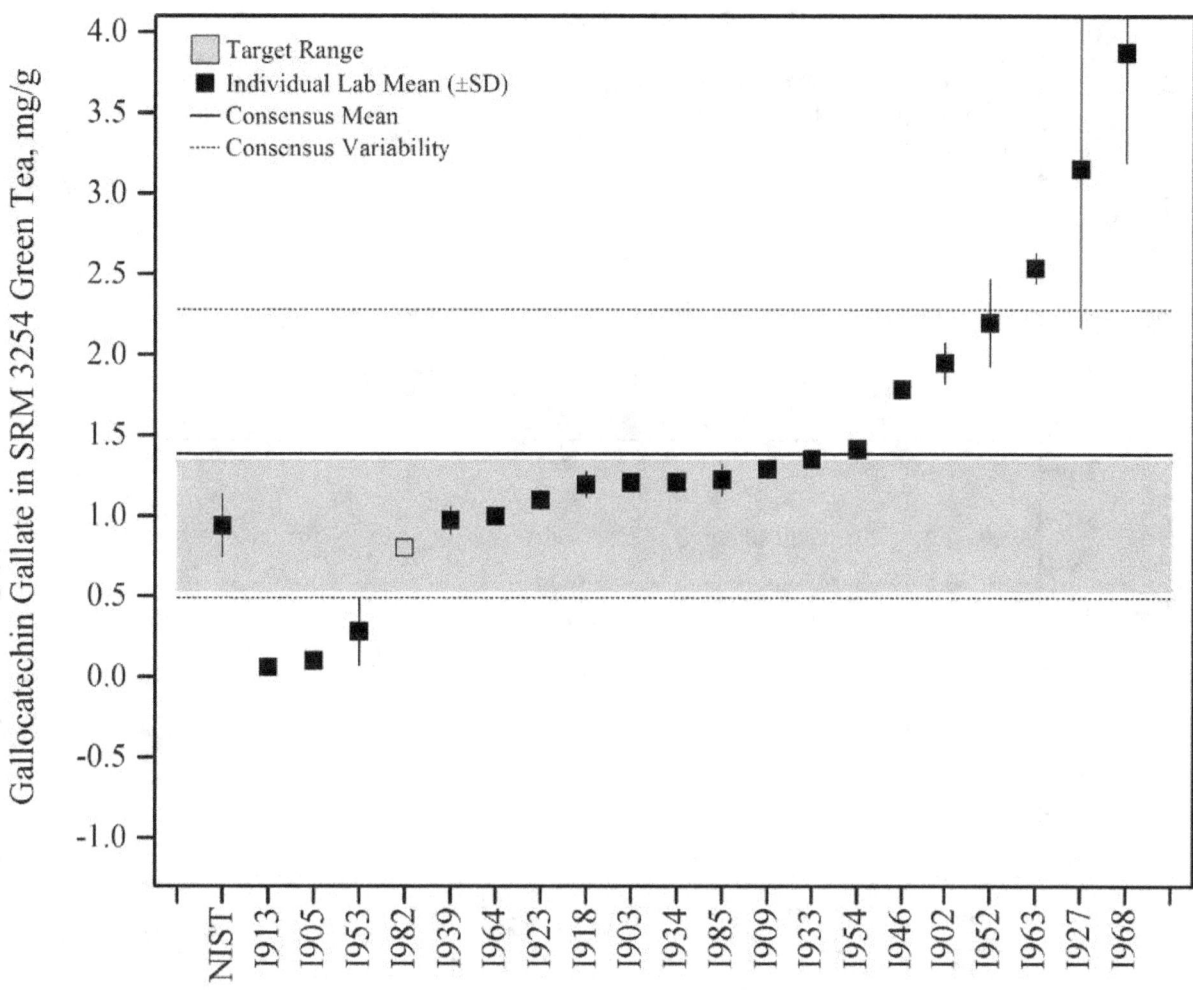

Figure 35. Gallocatechin gallate in SRM 3254 *Camellia sinensis* (Green Tea) (data summary view). In this view, individual laboratory data are plotted with the individual laboratory standard deviation (error bars). Data points that are unfilled represent laboratories that only reported a single value for that analyte and therefore were not included in the consensus mean. The black solid line represents the consensus mean, and the black dotted lines represent the consensus variability calculated as one standard deviation about the consensus mean. The gray shaded region represents the target zone for "acceptable" performance, which encompasses the NIST certified value bounded by twice its uncertainty (U_{95}).

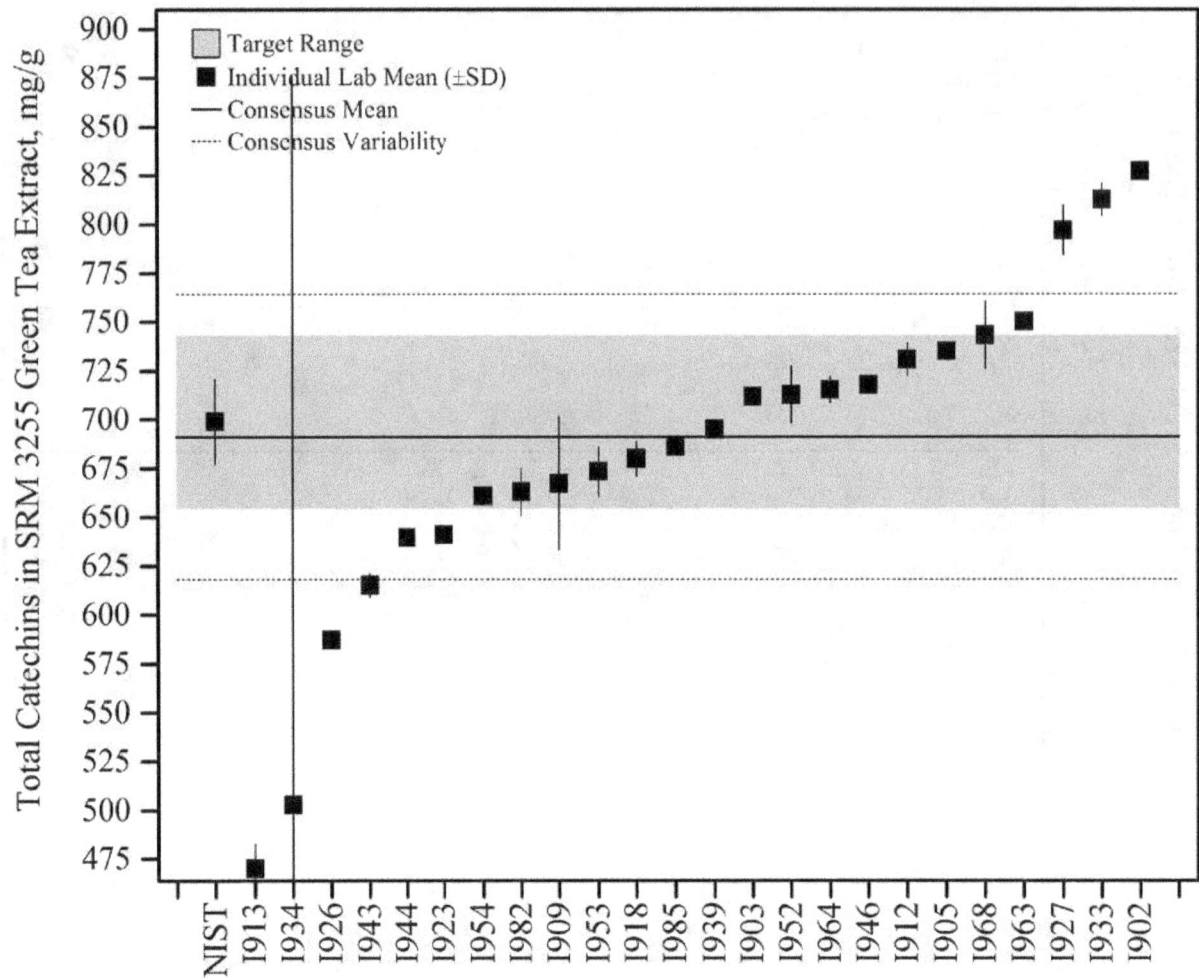

Figure 36. Total catechins in SRM 3255 *Camellia sinensis* (Green Tea) Extract (data summary view). In this view, individual laboratory data are plotted with the individual laboratory standard deviation (error bars). The black solid line represents the consensus mean, and the black dotted lines represent the consensus variability calculated as one standard deviation about the consensus mean. The gray shaded region represents the target zone for "acceptable" performance, which encompasses the NIST certified value bounded by twice its uncertainty (U_{95}).

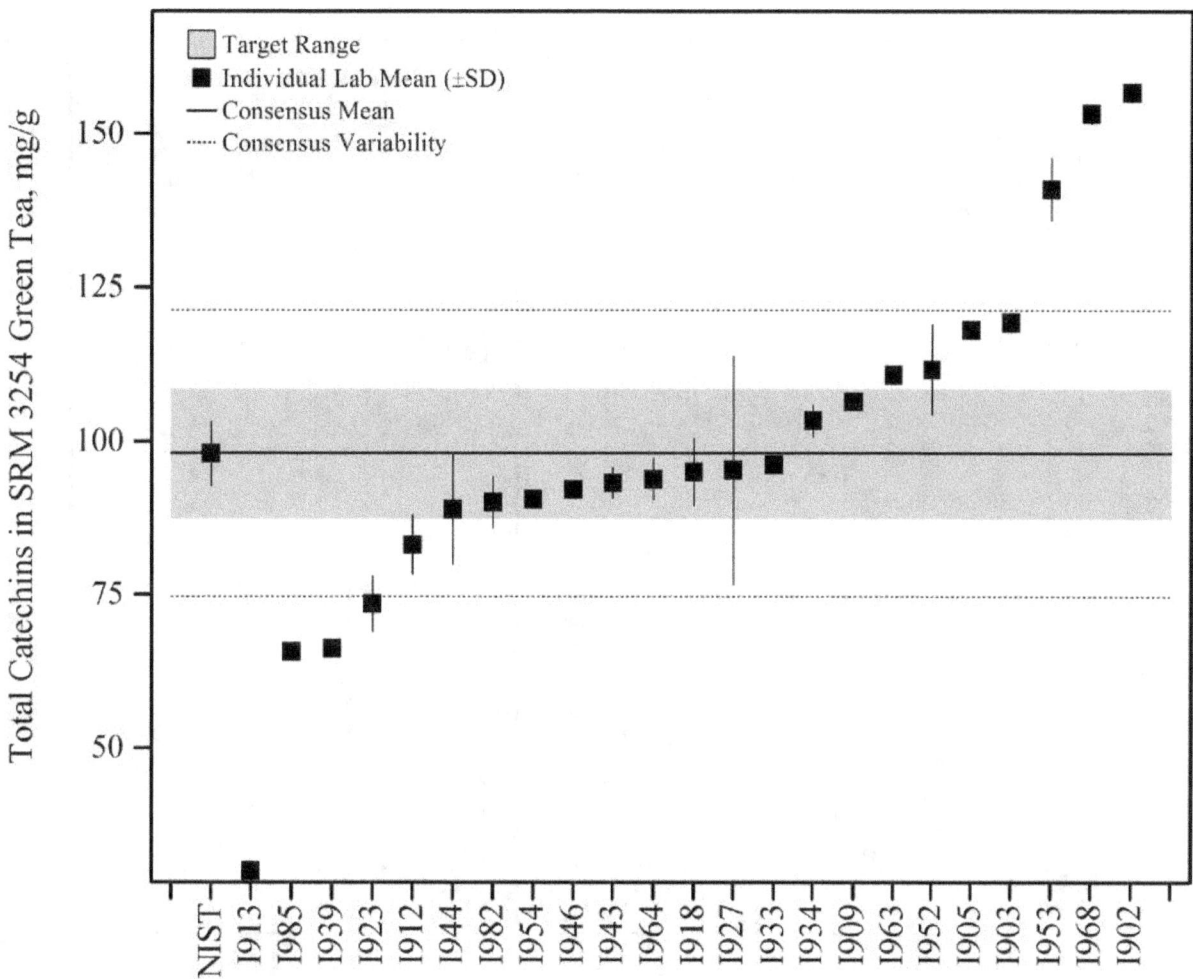

Figure 37. Total catechins in SRM 3254 *Camellia sinensis* (Green Tea) (data summary view). In this view, individual laboratory data are plotted with the individual laboratory standard deviation (error bars). The black solid line represents the consensus mean, and the black dotted lines represent the consensus variability calculated as one standard deviation about the consensus mean. The gray shaded region represents the target zone for "acceptable" performance, which encompasses the NIST certified value bounded by twice its uncertainty (U_{95}).

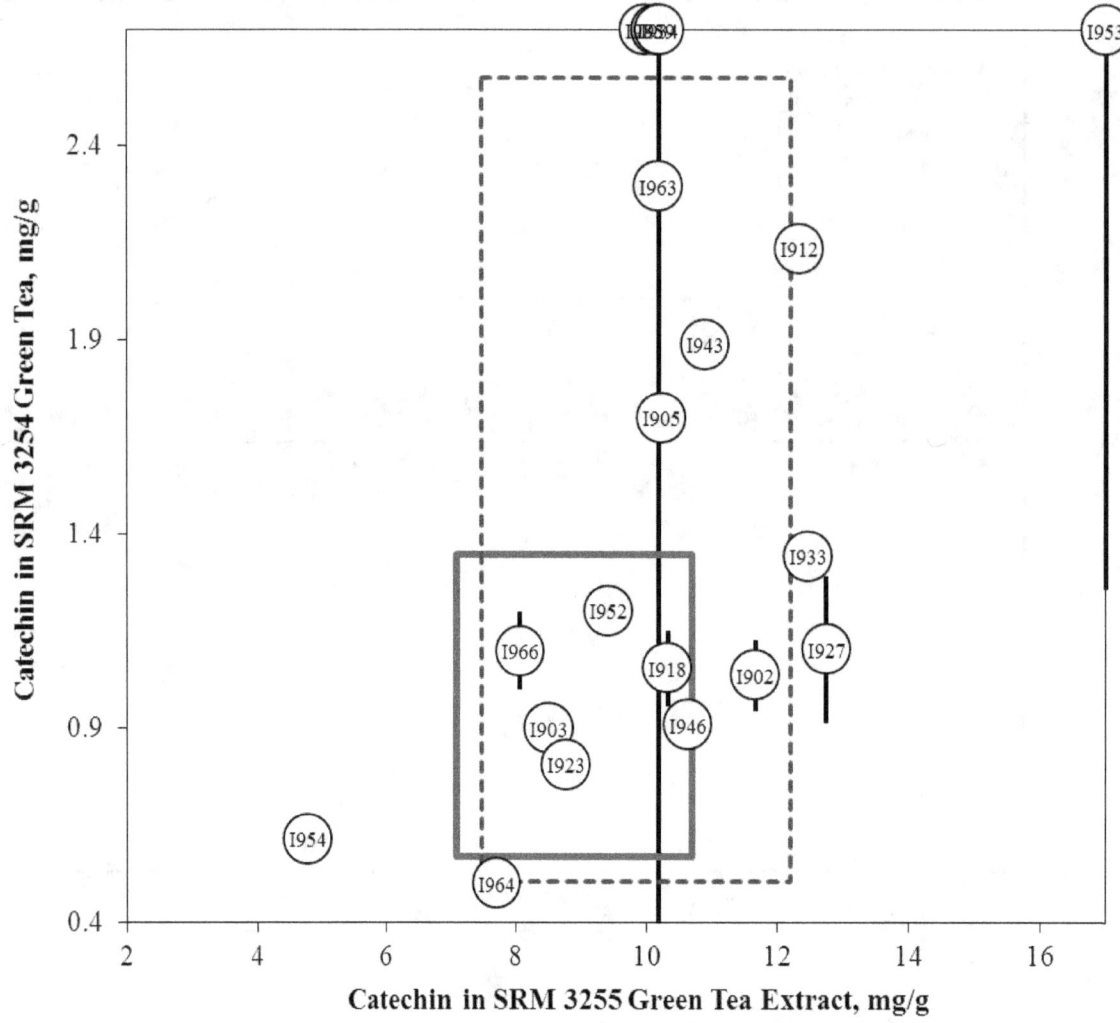

Figure 38. Catechin in SRM 3254 *Camellia sinensis* (Green Tea) Leaves and SRM 3255 *Camellia sinensis* (Green Tea) Extract (sample/control comparison view). In this view, the individual laboratory results for the control (SRM 3255 *Camellia sinensis* Extract) with a certified value for the analyte are compared to the results for an unknown (SRM 3254 *Camellia sinensis* Leaves). The error bars represent the individual laboratory standard deviation. The solid red lines represent the target zone for the control (x-axis) and the unknown sample (y-axis). The dotted blue box represents the consensus zone for the control (x-axis) and the unknown sample (y-axis).

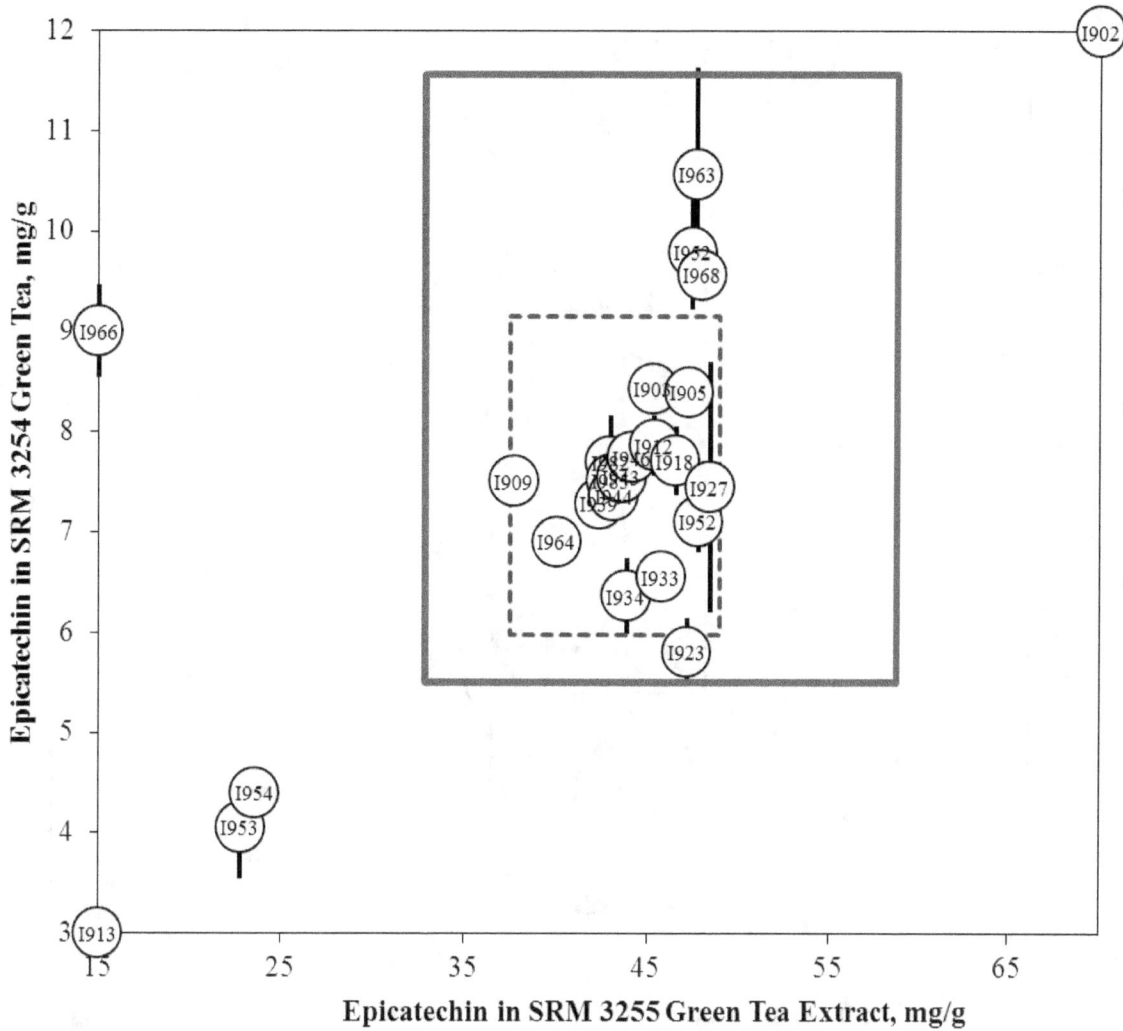

Figure 39. Epicatechin in SRM 3254 *Camellia sinensis* (Green Tea) Leaves and SRM 3255 *Camellia sinensis* (Green Tea) Extract (sample/control comparison view). In this view, the individual laboratory results for the control (SRM 3255 *Camellia sinensis* Extract) with a certified value for the analyte are compared to the results for an unknown (SRM 3254 *Camellia sinensis* Leaves). The error bars represent the individual laboratory standard deviation. The solid red lines represent the target zone for the control (x-axis) and the unknown sample (y-axis). The dotted blue box represents the consensus zone for the control (x-axis) and the unknown sample (y-axis).

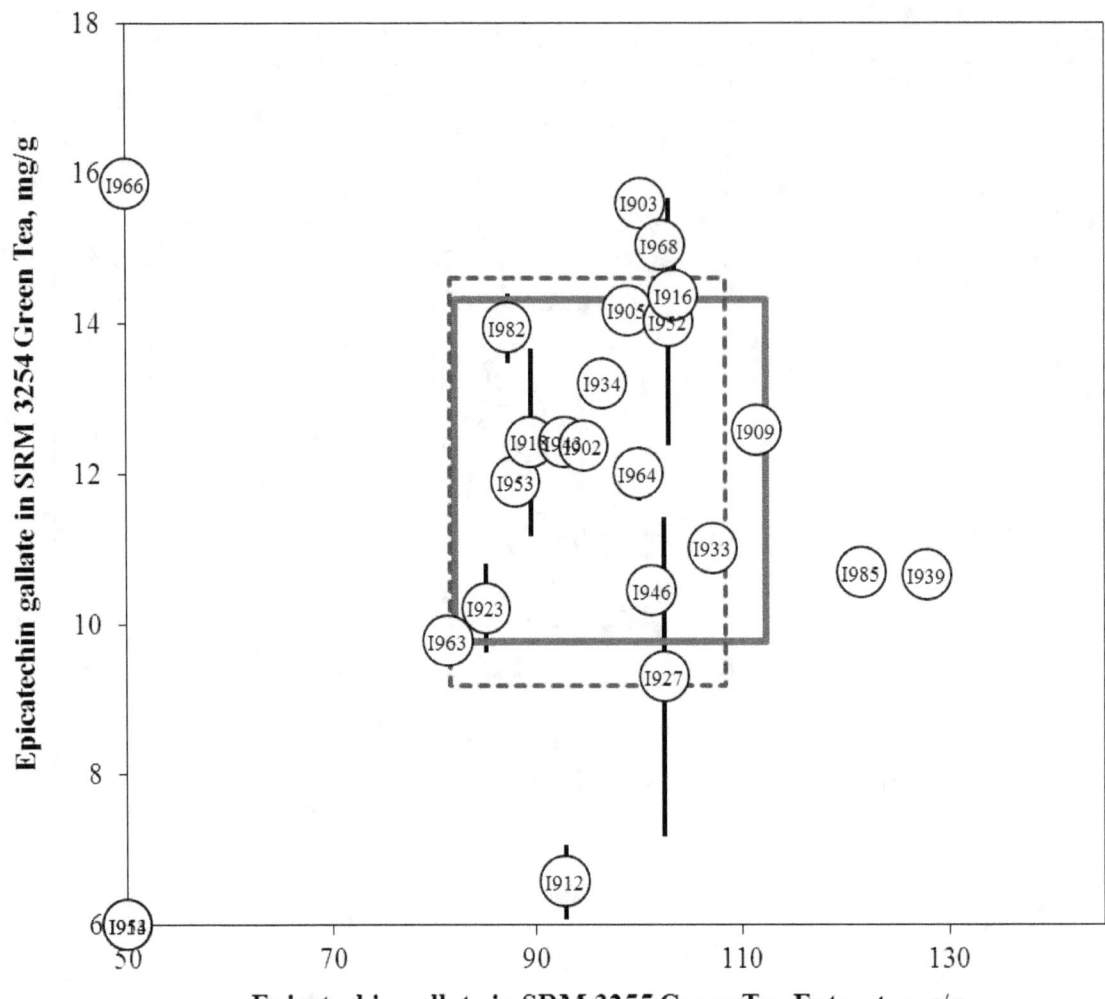

Figure 40. Epicatechin gallate in SRM 3254 *Camellia sinensis* (Green Tea) Leaves and SRM 3255 *Camellia sinensis* (Green Tea) Extract (sample/control comparison view). In this view, the individual laboratory results for the control (SRM 3255 *Camellia sinensis* Extract) with a certified value for the analyte are compared to the results for an unknown (SRM 3254 *Camellia sinensis* Leaves). The error bars represent the individual laboratory standard deviation. The solid red lines represent the target zone for the control (x-axis) and the unknown sample (y-axis). The dotted blue box represents the consensus zone for the control (x-axis) and the unknown sample (y-axis).

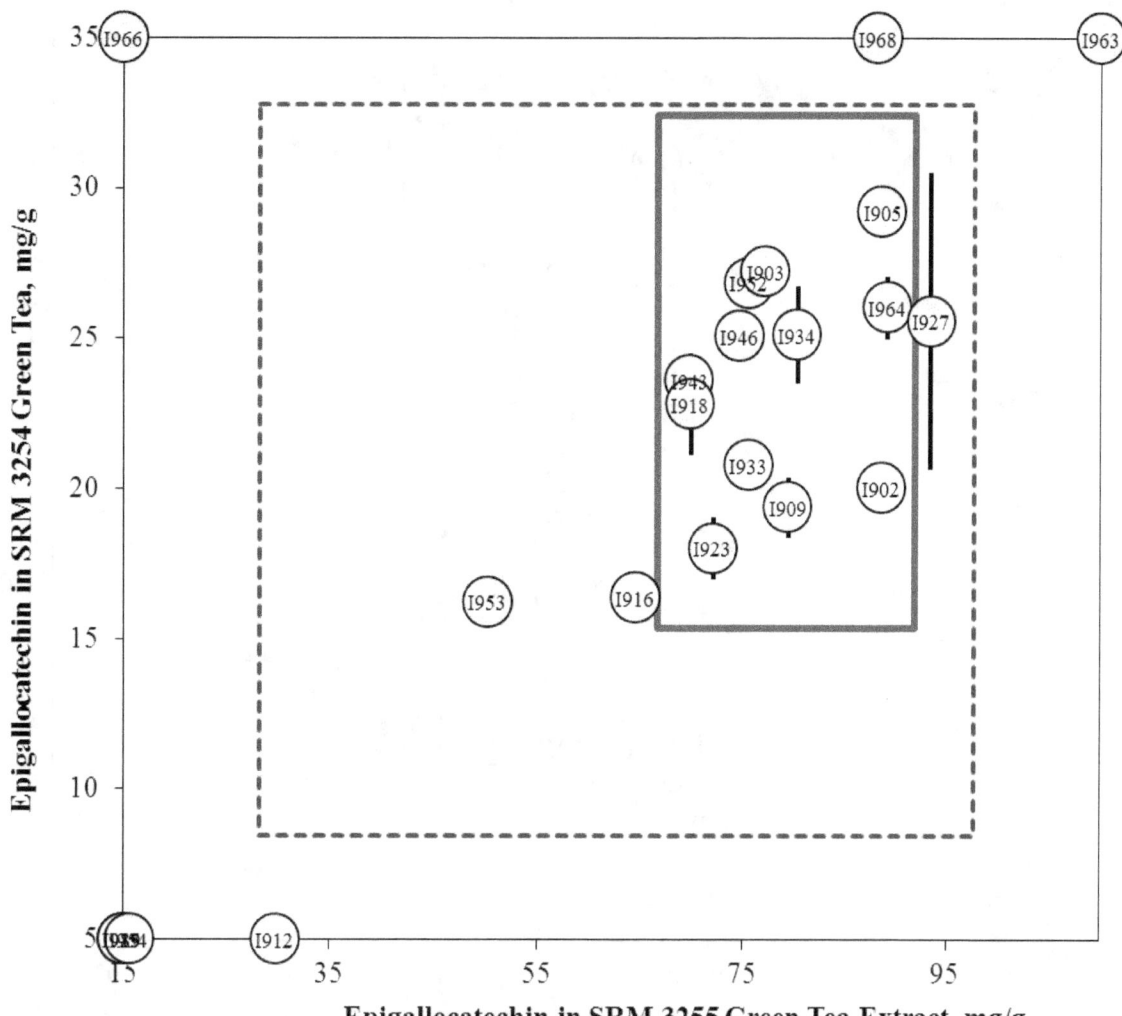

Figure 41. Epigallocatechin in SRM 3254 *Camellia sinensis* (Green Tea) Leaves and SRM 3255 *Camellia sinensis* (Green Tea) Extract (sample/control comparison view). In this view, the individual laboratory results for the control (SRM 3255 *Camellia sinensis* Extract) with a certified value for the analyte are compared to the results for an unknown (SRM 3254 *Camellia sinensis* Leaves). The error bars represent the individual laboratory standard deviation. The solid red lines represent the target zone for the control (x-axis) and the unknown sample (y-axis). The dotted blue box represents the consensus zone for the control (x-axis) and the unknown sample (y-axis).

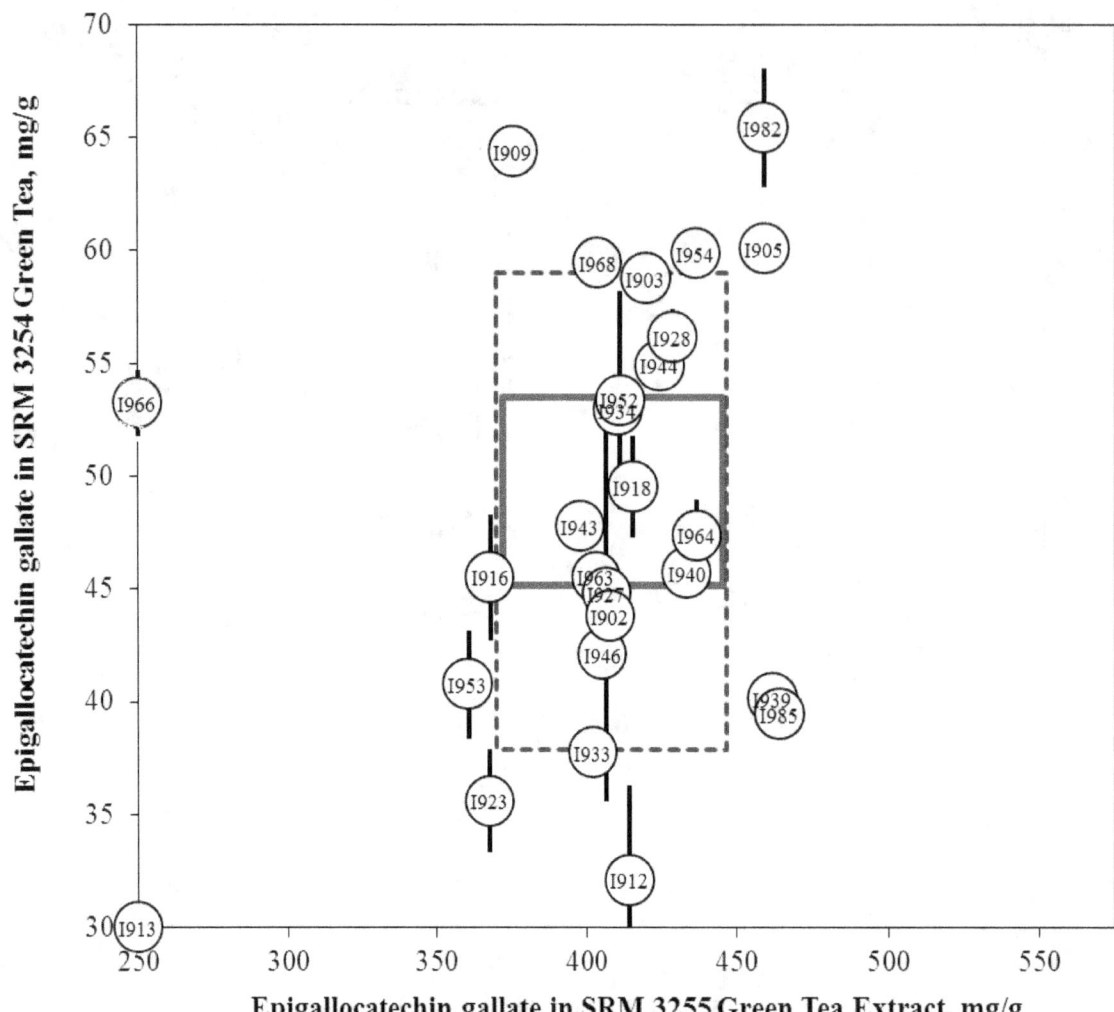

Figure 42. Epigallocatechin gallate in SRM 3254 *Camellia sinensis* (Green Tea) Leaves and SRM 3255 *Camellia sinensis* (Green Tea) Extract (sample/control comparison view). In this view, the individual laboratory results for the control (SRM 3255 *Camellia sinensis* Extract) with a certified value for the analyte are compared to the results for an unknown (SRM 3254 *Camellia sinensis* Leaves). The error bars represent the individual laboratory standard deviation. The solid red lines represent the target zone for the control (x-axis) and the unknown sample (y-axis). The dotted blue box represents the consensus zone for the control (x-axis) and the unknown sample (y-axis).

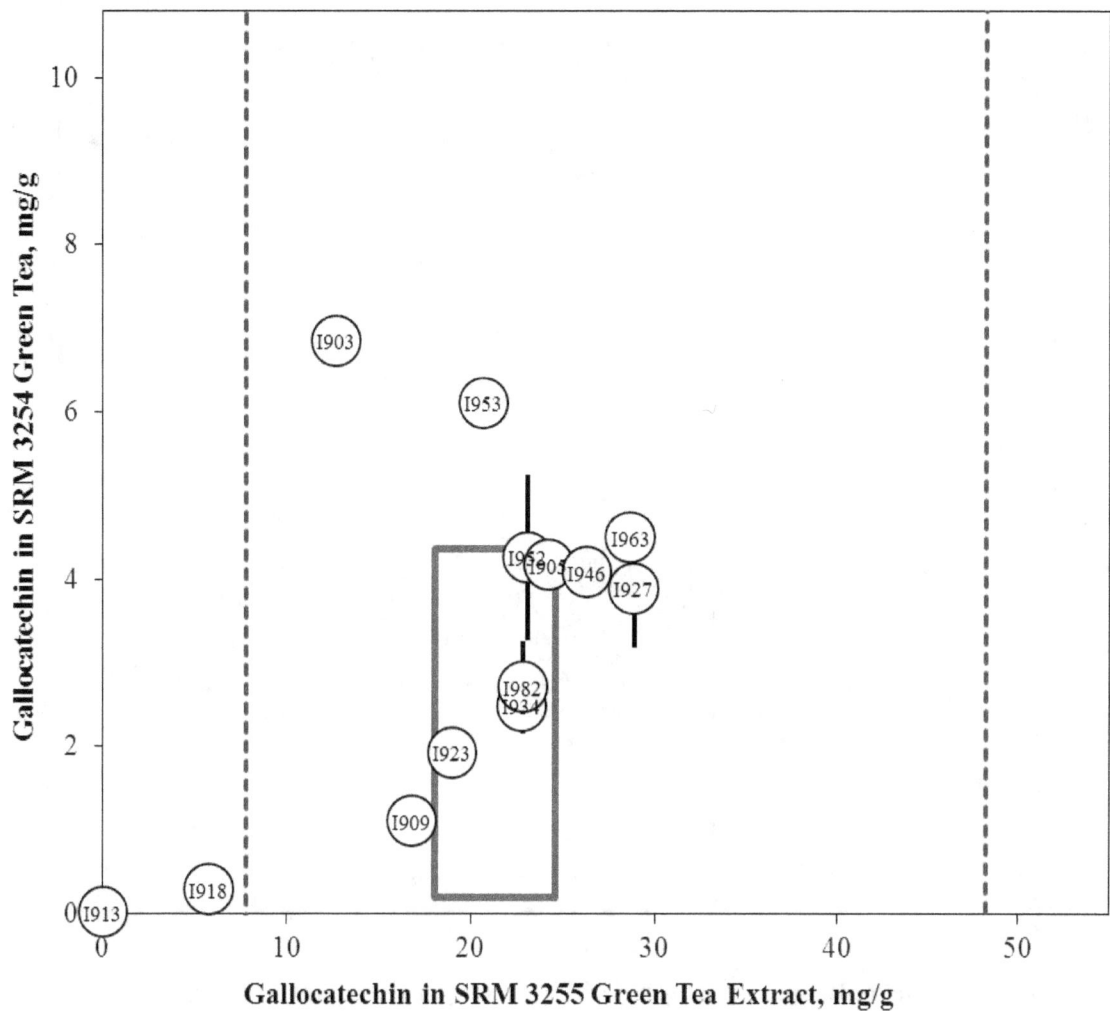

Figure 43. Gallocatechin in SRM 3254 *Camellia sinensis* (Green Tea) Leaves and SRM 3255 *Camellia sinensis* (Green Tea) Extract (sample/control comparison view). In this view, the individual laboratory results for the control (SRM 3255 *Camellia sinensis* Extract) with a certified value for the analyte are compared to the results for an unknown (SRM 3254 *Camellia sinensis* Leaves). The error bars represent the individual laboratory standard deviation. The solid red lines represent the target zone for the control (x-axis) and the unknown sample (y-axis). The dotted blue box represents the consensus zone for the control (x-axis) and the unknown sample (y-axis).

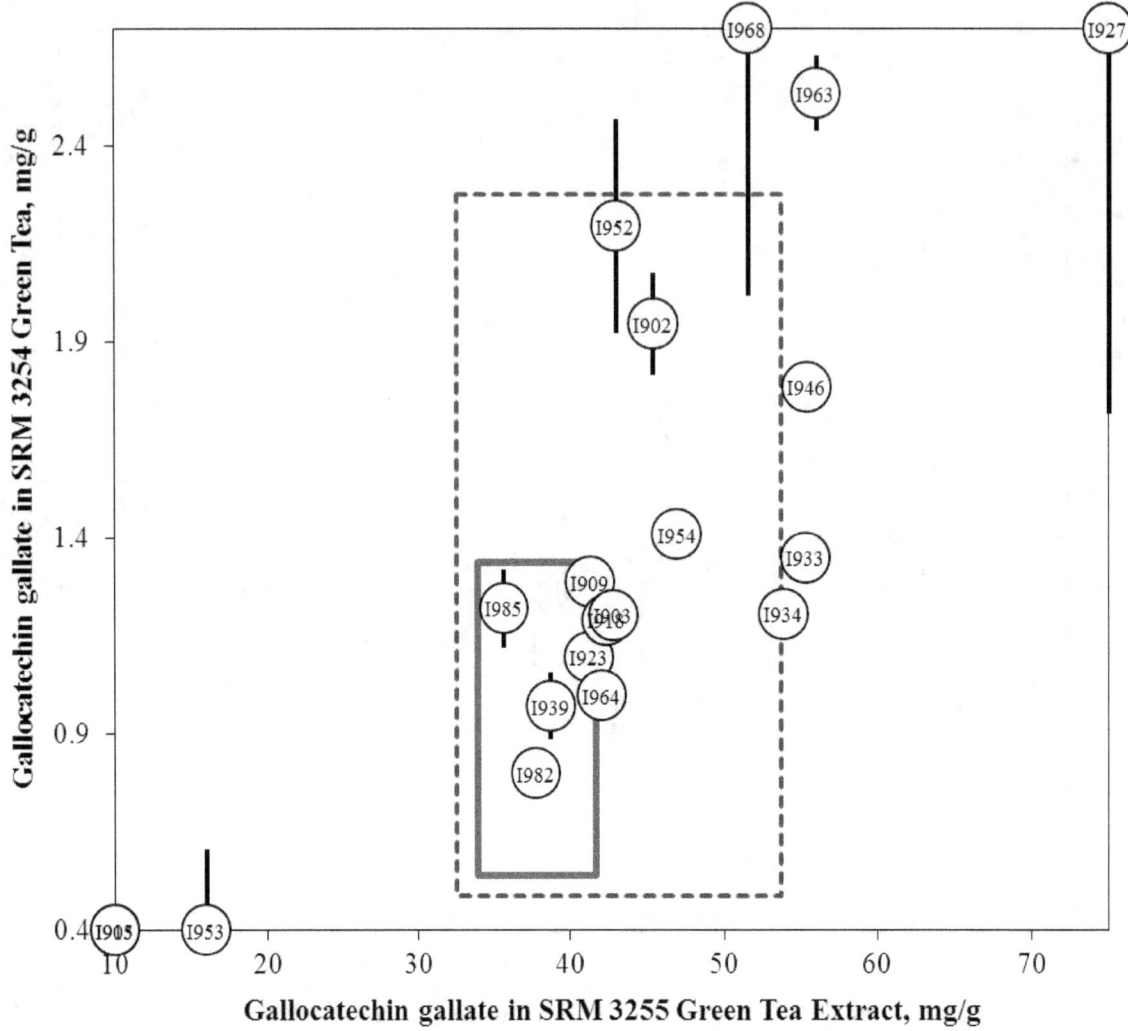

Figure 44. Gallocatechin gallate in SRM 3254 *Camellia sinensis* (Green Tea) Leaves and SRM 3255 *Camellia sinensis* (Green Tea) Extract (sample/control comparison view). In this view, the individual laboratory results for the control (SRM 3255 *Camellia sinensis* Extract) with a certified value for the analyte are compared to the results for an unknown (SRM 3254 *Camellia sinensis* Leaves). The error bars represent the individual laboratory standard deviation. The solid red lines represent the target zone for the control (x-axis) and the unknown sample (y-axis). The dotted blue box represents the consensus zone for the control (x-axis) and the unknown sample (y-axis).

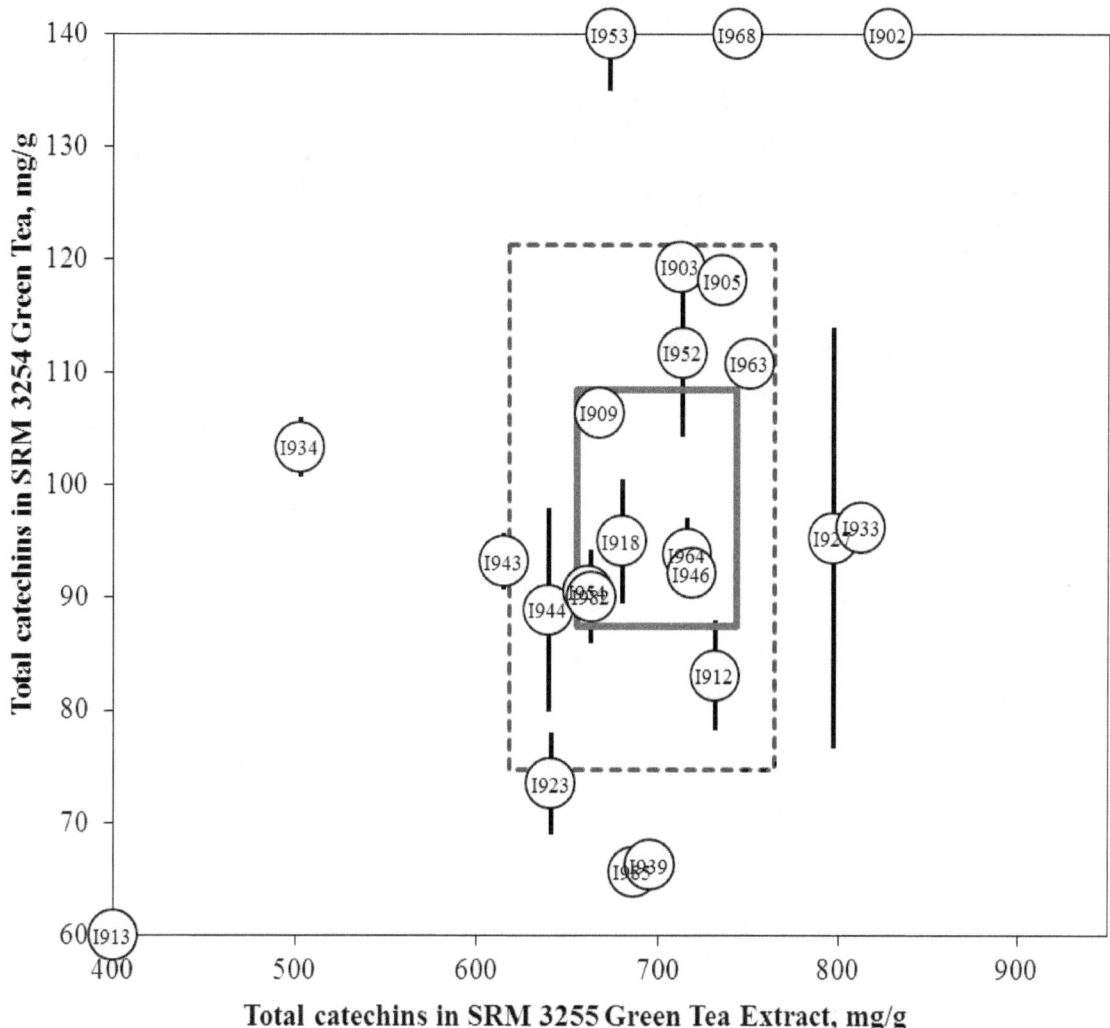

Figure 45. Total catechins in SRM 3254 *Camellia sinensis* (Green Tea) Leaves and SRM 3255 *Camellia sinensis* (Green Tea) Extract (sample/control comparison view). In this view, the individual laboratory results for the control (SRM 3255 *Camellia sinensis* Extract) with a certified value for the analyte are compared to the results for an unknown (SRM 3254 *Camellia sinensis* Leaves). The error bars represent the individual laboratory standard deviation. The solid red lines represent the target zone for the control (x-axis) and the unknown sample (y-axis). The dotted blue box represents the consensus zone for the control (x-axis) and the unknown sample (y-axis).